Advance Praise

This is a powerful book on how today's enterprise can make personalized content happen. Use it wisely and succeed while your competitors flounder.

—Joe Pulizzi, Founder, The Tilt and Content Marketing Institute

I love love LOVE this book. It puts the focus—finally!—on the lynchpin of any truly effective personalized online experience: well-structured, well-prepared content. The result? Actionable, useful content when, where, and how your audience needs it. Insightful, instructive, and a highly enjoyable read. Everyone needs this book, now.

—Kristina Halvorson, CEO and Founder, Brain Traffic. Author of *Content Strategy for the Web*

Finally, a clear, concise, and common sense approach to creating personalized content experiences at scale. Solid advice for now and into the future. A must-read for any content professional.

—Scott Abel, The Content Wrangler

Val and Regina have finally written the long-overdue manual on how to actually realize the potential of personalized content. Following their advice is a sure roadmap to success!

—Leslie Farinella, Chief Operating Officer, Xyleme

Many industry experts have written about the importance of personalization, but this is the first book that really lays out the map of how to get there. The authors give you the hard truths, busting more than a few big myths along the way. It's a realistic look, born from decades of experience, at what you'll need to consider to make the magic happen.

—Rahel Anne Bailie, Founding CEO, Content, Seriously Consulting

Val's new book is a "how to guide" to enterprise-grade content personalization strategy. *The Personalization Paradox* should be required reading for anyone involved in the digital customer experience value chain.

—Shane Cumming, Chief Revenue Officer, Acrolinx

Is your company seeking the holy grail of content personalization? You're in for a wild ride. The adventure begins with an apparent paradox: standardizing your content. With this book, Val and Regina lead the way.

—Marcia Reifer-Johnston, Author of *Word Up! How to Write Powerful Sentences and Paragraphs (And Everything You Build from Them)*

If you're looking to communicate in the new online world, this book is a must. Today, effective writing requires so much more. In this book Val and Regina break the components down to basics, and show you in easy steps what you need to do to communicate, and avoid those errors you're seeing every day.

—Mark Gross, CEO of Data Conversion Laboratory

Val and Regina have created an important reference book. No relevant aspect is left unexamined. A must-read for everyone involved in content creation!

—Stefan Kreckwitz, CEO of Congree Language Technologies

The Personalization Paradox is required reading for anyone responsible for delivering personalized information to a customer, employee or partner. Val and Regina skillfully map, chapter by chapter, a process to develop a strategy and plan. The result is an ingenious process of how to navigate the maze of requirements to develop optimal content, metadata and processes. As Charles Cooper states, this is not a technological change. This is a cultural change. Val and Regina present a fabulous methodology that is compelling and adaptable.

—Chip Gettinger, VP Global Solutions Consulting, RWS

The Personalization Paradox

Why Companies Fail (and How To Succeed) at Delivering Personalized Experiences at Scale

Val Swisher

Regina Lynn Preciado

The Personalization Paradox

Why Companies Fail (and How To Succeed) at Delivering Personalized Experiences at Scale

Credits

Cover design and illustrations	Kiam Jamrog-McQuaid
Editor	Lisa Péré
Managing Editor	Max Swisher

Disclaimer

Trademarks

XML Press
Laguna Hills, California 92637
http://xmlpress.net

First Edition
978-1-937434-72-4 (print)
978-1-937434-73-1 (ebook)

Table of Contents

Welcome to the Paradox

By Robert Rose, Author,
Chief Strategy Officer, The Content Advisory

Research shows that surprise can intensify our emotions by about 400 percent.[1] It's why something as simple as a box of chocolates you didn't expect can make your day, or why a careless jerk who steals your parking space can ruin it.

But what about surprises that you expect?

Perhaps it's an unopened present, the new job you're starting Monday, or the next episode of that amazing Netflix show. You know you'll be surprised. But you have no idea *how* you'll be delighted.

Here's a paradox: The anticipation of opening a present from your favorite uncle can be more exciting than the present itself. The hopeful expectation of an over-the-top-fantastic beach vacation can be more happy-making than the trip.

Isn't it odd, then, that in modern digital business, we believe that removing all the expected surprise from our customer experiences somehow makes sense? Businesses see technology that can deliver personalized, targeted content as the most optimal means of giving customers exactly what they *expect* to see. And yet, most businesses have no clue how to create or structure that content to provide any level of delightful *surprise* in that experience.

So, what's the result? Most personalized content initiatives consist of a distinctly *impersonal* variation of "Hi [*FirstName*], we know you showed interest in [*ProductViewed*], would you like to buy it for [*TargetedDiscount*]?" That isn't personalization. And it definitely isn't an expected surprise. It's simply a new iteration of the same old experience.

And that's a problem for marketers. As Ben Hoff put it in *The Tao of Pooh*, "Each time the goal is reached, it becomes not so much fun, and we're off to reach the next one, then the next one, then the next."

---

[1] "The Takeaway: Surprise! Why the Unexpected Feels Good, and Why It's Good For Us" (WNYC 2015)

And in today's world, emergent privacy concerns and data-usage regulations create an even bigger challenge. It's yet another paradox: Ask customers if they want more targeted, relevant advertising and content as part of the buying process, and most will say, "Yes!" But ask them if companies should use their data to deliver that content, and they will overwhelmingly say, "Hell no!"

What excites me most about the book you are about to read is that it deftly and simply demystifies one of the biggest challenges in today's modern marketing and customer experience development: How can businesses truly deliver the expected surprise of content personalization at scale?

Val and Regina go well beyond the standard "solutions" (more data, time, or technology) to provide an in-depth road map of the only path toward personalizing at scale: standardization. They walk you through a content re-use strategy and explain why it's integral. And maybe most important, they illuminate what personalization *really* looks like, at every level: word, sentence, paragraph, and page.

The best content experiences aren't those that are conspicuously personalized by throwing your name or some other personal data in your face to give you a "familiar" experience. They are experiences that seem to know you well enough to surprise you.

As a modern marketer, the best compliment you can get from a customer isn't that your content met their expectation. No, the best compliment is when someone says, "I really look forward to the next …."

The Personalization Paradox can help you deliver delightful, personalized digital experiences at scale. It can help you not only deliver a positive outcome for your customers but also intensify their expectation of delight.

I know you don't know what comes next in the following pages. But I promise you, it's going to be great. Prepare for an expected surprise.

What Is the Personalization Paradox?

Content personalization has become the ambition of modern communications. Over the span of just a few years, we've seen the ability to deliver personalized experiences change from "nice to have for marketing efforts" to "a business imperative for content across the company."

Marketing, communications, human resources, training, technical documentation, and customer support always aim to deliver relevant, usable, and timely content. They want to deliver:

- The right content
- To the right person
- At the right time
- On the right device
- In the language their choosing

But to stay competitive, companies of all sizes now need to deliver the specific content each customer needs—nothing more and nothing less—when, where, and how the customer needs it. And they need to do it for *every* customer. This is personalization at scale.

Two Ways to Provide Personalization

There are two ways to personalize content: manual and automated.

In **manual personalization**, you create, manage, store, update, and retire different content for each person, persona, or customer type. Many companies have tried—and failed—to deliver personalized content in this way. Not only does creating personas stereotype content consumers in a way that limits the effectiveness of your content, but this approach simply doesn't scale.

Automated personalization emphasizes sophisticated tools that attempt to match content to consumer. In automated personalization, you store content assets in a content management system (CMS), then mix and match appropriate pieces of content to create personalized output at the point of delivery.

CMS vendors like Sitecore, Ektron, Magnolia, and Episerver have offered automated personalization features for more than a decade. Many companies have deployed expensive new software and high-powered analytics to deliver personalization proofs of concept, using a limited set of content and customer data. Yet for all the promise of personalized content and the ROI it is supposed to deliver, we cannot point to a single company that is successfully delivering personalization for actual content across the company.

Our industry has been trying to achieve the ultimate goal of personalization at scale for a long time. So why aren't we succeeding?

Why Does Personalization Fail?

Personalization is the sexiest thing that no one is doing.
— Robert Rose, Content Strategy and Customer Experience Expert

There are two overarching reasons behind the failure of personalized content at scale. The first (and the one on which most companies focus their efforts) is difficulty identifying which information a content consumer wants. For years, we either played a guessing game or approached consumer needs from an elite, corporate viewpoint.

Big Data has largely solved this problem. Social media and search companies have become extremely good at keeping track of us. Most of us willingly provide a huge amount of information to companies that then collect and sell it. Although uncomfortable (some even say immoral or illegal), the fact is that companies of all sizes purchase this information so that they can target the content you see based on the data that has been collected about you.

The second reason that personalization fails is a complete lack of focus on the content itself. To date, few companies have optimized their content for reuse, automation, or personalization.

Instead:

- They focus on content delivery rather than on content creation and management.
- They invest heavily in new tools and systems, but do not transform their content.
- They do not approach the process holistically, instead trapping content in silos.

The Secret of Successful Personalization at Scale

The only way to deliver personalized content at scale is to automate the content process at the point of delivery. But for that approach to work, you must change how you *do* content.

Historically, we were taught to create content by starting at "page 1" and writing to the end of the asset. The result was typically a large document or a collection of large documents. This method, which mandates that we create a different long-form asset for every personalized version that we need, is not scalable.

To personalize at scale, we need to change the paradigm of how we create, store, manage, and retire content. Instead of creating long-form content from beginning to end, we need to create and use small, nimble chunks of content. We must then enable those chunks to be automatically combined to create the experience a customer needs, when they need it. This personalized experience can be:

- A months-long digital journey from prospect to customer
- Record-time product setup, thanks to documentation that includes only the needed instructions
- Successful completion of an e-learning module
- A five-minute problem solving stop on a support website
- Review of regulatory material to verify that everything is in place, in order, and in compliance

To create nimble, reusable pieces of content that can be combined in different ways for different people and different devices, you *must* standardize everything about that content:

- The words and images you use
- The ways in which you use those words
- The paragraphs and sets of paragraphs
- The overall tone and voice

If you do not standardize your content, you cannot successfully combine various chunks of content in different ways. Without standardization, your result will be a "Frankendoc" that might not make sense to your customer. Personalization without standardization is a recipe for a confusing, disjointed customer experience.

And herein lies the Personalization Paradox: **To provide personalized experiences at scale, the content itself must be standardized.**

The Paradox: Standardization Enables Personalization

Standardizing content to create a personalized experience might seem counterintuitive at first. After all, when we think of a personalized experience, we think of unique content created for and delivered to a unique individual. But creating unique content in this way simply does not scale.

Without standardization, any attempt to deliver a personalized experience is hampered by content that does not flow when the consumer encounters it. The experience sends mixed messages. The content creates confusion instead of providing clarity.

By using standardization, your content can mix and match seamlessly. Content components are uniform in terminology, tone, grammar, and aesthetic. They are tagged with rich metadata, so that systems and people can find them. And components are stored in a CMS that makes content easy to find, assemble, and release to your personalization engine and delivery platforms.

Who Can Benefit from Reading This Book?

Our goal is to give you a solid plan for successfully personalizing content at scale, without getting lost in minutiae. We've tried to strike the right balance between the forest and the trees.

The Personalization Paradox: Why Companies Fail (and How to Succeed) at Delivering Personalized Experiences at Scale describes what you need to do and why. The book reveals the missing piece that causes so many companies to struggle. By focusing on how to standardize content, we show you how to create content that you can use to deliver personalized experiences at scale.

We wrote this book for anyone who manages content teams—even a team of one. If you or your team produces content for …

- Marketing or digital marketing
- Product or technical documentation
- Customer or technical support
- Training and education
- Sales enablement
- Internal or external communications

… then this book is for you.

CHAPTER 1

The Content Conundrum

Creating personalized experiences at scale presents a content conundrum for almost every organization. Why?

When you use the same old content, written the same old way via the same old tools, the only way to personalize the consumer experience is to create separate content sets for each individual. Using existing methodologies, the best you can do is copy, paste, and tweak each piece of content to create a unique output for each possible person. That leaves you with an enormous amount of content to create, manage, update, and deliver. You also face the daunting task of managing all the data that accompanies the content. It quickly becomes clear why personalization using old content, tools, and methods doesn't scale.

Delivering personalized content at scale requires an altogether different content ecosystem.

Tools alone—even the shiniest new ones—won't solve the problem. In recent years, we've seen many new and exciting tools try to solve the conundrum of matching content to customer at the time of delivery. What we have *not* seen is a clear understanding of how the content itself needs to change. You cannot simply use new tools to personalize the same content you've been creating in the same old way.

This is the mistake that almost every company is making. They are tossing old content into new tools and expecting those tools to completely solve the problem.

But to successfully personalize content at scale, you must rethink the content itself.

I think that the true gem is anticipation of people's needs. That's why you're personalizing content, right? Because you're trying to anticipate which content will actually move them to the next stage, towards a buy.
—Marcus Hearne, SVP, Product & Solutions Management

Easy-to-Find Information Is Not New

Automated content personalization is a technical outgrowth of the venerable desire to make content easier to find. The concept of *findability* is not new; it has always been a key factor in content creation. After all, what good is information if you can't locate it?

For centuries, content creators have grappled with how to make information easy to locate. A table of contents, a list of figures, list of tables, and an index are all examples of navigation techniques long used in print publications.

- According to the American Society for Indexing, one of the first instances of a large table of contents was created by Pliny the Elder (AD 23/24–79) in *Naturalis Historia* (Pliny 1669). This enormous tome is organized into 37 books in 10 volumes. The first book is nothing more than a table of contents to help readers locate information in the other 36 books.[1]
- The first known English index was created in 1575 for *A booke of the arte and maner how to plant and graffe all sortes of trees* (Mascall 1575) by Leonard Mascall.[2]
- The card catalog (those little index cards stored alphabetically in multi-drawer wooden cabinets in the library) was invented in the late 1700s.[3]

Of course, before modern technology, the critical work required to create a table of contents or an index was arduous. The invention of technical publishing software was key to automating the process of creating navigation like a table of contents or index. Automating the way we create navigation tools was a critical step in the history of findability (and a great use of technology).

Take FrameMaker—the first technical publishing software that Val ever used. In the late 1980s, FrameMaker made it easy to generate a table of contents, lists of figures and tables, and an index. At about the same time, content creators began using similar technology to create cross-references and heading levels. These features helped us organize information and provided content consumers with guideposts for finding the information they needed. And they still do today.

In the early 1990s, the internet became widely available to the public. Shortly after, content creators and user-interface designers added website navigation and online help systems to the growing list of ways to help people find the information they needed as quickly and painlessly as possible.

---

[1] *History of Information Retrieval* (American Society for Indexing)

[2] "Index (Publishing)" (Wikipedia)

[3] *The Card Catalog* (Hayden 2017)

Pulling Information

From Pliny the Elder's table of contents to website navigation, these findability methods have one thing in common: They require the content consumer to search for information. The onus is on the consumer to look up the information or click the right button.

We call this type of technology, in which a person has to pull the information out of a printed book or a digital repository, *pull technology*.

The problem with pull technology is that it can be extremely time-consuming. It can also be unreliable. We all know the frustration of getting too many search results. *Will the real answer to my question please make itself known?*

Pushing Information

Companies spent years looking for a better way to provide people with the information they need, when they need it. Rather than making the content consumer hunt for the information, we developed a way to remove friction and directly provide the right information. We call this *push technology*: providing the information the consumer needs with little or no action on their part.

Push technology needs two things to be successful:

- Accurate data about the content a customer needs at a specific moment
- Accurate content delivered in the right format and the desired language

Much has been written about the first need: collecting and analyzing accurate data about the content someone needs at a particular moment. Historically, predicting the content someone needed was a roadblock to successfully delivering a personalized experience. We simply didn't have enough analytic information to be predictive.

Today, predictive technology based on Big Data analytics is at the heart of many marketing campaigns. Companies have been able to collect enormous amounts of data on our online behaviors, desires, and preferences. These same companies readily sell that data to other companies that are willing to pay handsome sums of money for the information. Thanks to Big Data, figuring out the right content for a particular person at a particular moment isn't nearly the challenge it was just a decade ago.

It's the second need—the ability to deliver accurate content (and *only* that content) in the right format and language—that continues to elude companies both large and small.

Why Personalize?

Content personalization is nothing more than the next step in content findability. The main reason to personalize experiences is to make it easy for a customer to find the information they need when they need it: The Holy Grail of customer interaction.

From a corporate perspective, one of the most important reasons to provide personalized experiences is the relationship between personalized content and sales. Study after study reports a direct correlation between personalized experiences and increased sales.

Personalized experiences increase brand loyalty. Customers who receive personalized content feel that the company really understands them. These customers tend to become repeat customers.

By personalizing content, we are trying to reduce or eliminate *resistance*—the amount of work that customers must do to receive their desired goods or services. We can measure customer resistance in several ways:

- The number of clicks it takes for customers to find the information they are looking for
- The proportion of text and imagery on a page that is relevant to the customer
- How far down the page customers must scroll to find the information they need
- How many interactions customers must have with a chatbot before they get the answer to their question
- How many search queries customers attempt before finding the right result

By personalizing content, we lower resistance to as small a value as possible.

Something of paramount importance, which drives an increased need to be personal, is that the customer experience is increasingly becoming the most important thing in terms of brand differentiation. To create this type of personal experience, a lot goes into thinking about the way everything connects and also to creating a seamless omnichannel experience. You can put all the pieces together, but if you don't communicate in a seamless and personal way across the different channels, you're not delivering the kind of ideal experience that will help you differentiate as a brand.

—Harvey Turner, COO, Pilot44

Personalization by the Numbers

How effective is content personalization? Many studies focus on the answer. Looking at the data, two things stand out:

- Most companies say that personalizing the customer experience is a critical "must have," and they have the statistics to back it up.
- Few companies believe they are delivering enough personalized content or delivering it well.

According to Salesforce, almost 60 percent of customers feel that a personalized experience is very important when they make a buying decision. In addition, the fifth edition of the Salesforce State of Marketing research report credits personalization for improving:[4]

- Brand building
- Lead generation
- Customer acquisition
- Upselling
- Customer retention
- Customer advocacy

The 2019 Trends in Personalization study by Evergage (a Salesforce company) and Researchscape International provides a number of compelling statistics:[5]

- Almost three-quarters of surveyed marketers report a sales increase of more than 10 percent due to personalizing content.
- Only 2 percent of marketers feel that personalization has no impact on customer relationships, whereas more than 70 percent report that personalization has either a strong or extremely strong impact.
- A whopping 85 percent of marketers believe their customers expect a personalized experience.
- Less than one-third of marketers believe they are successful in their personalization efforts, and only 16 percent of respondents are very satisfied or extremely satisfied with the level of personalization they are providing.

---

[4] *State of Marketing* (Salesforce Research 2019)

[5] *2019 Trends in Personalization* (Evergage 2019)

According to Infosys, 86 percent of consumers surveyed indicate that personalized content has some impact on what they purchase, and 25 percent say that personalization plays a large role in their purchases. However, in the same Infosys study, 20 percent of respondents say they have never interacted with personalized content, and about one-third say they want to see more personalization than they currently experience.[6]

According to a study commissioned by Monetate, "More than 70 percent of those [companies] that reported profitability increases in the last 12 months have a documented personalization strategy, compared to less than 20 percent for those who reported decreased profitability."[7]

How to Personalize at Scale

How can you create a unique personalized experience for each person and deliver these experiences at scale? There is only one way: Break up monolithic pieces of content into small, nimble, reusable components.

Quite simply, we have the capability to deliver personalized experiences—
if content is segmented and tagged, end to end.
—Charles Cooper, Vice President, The Rockley Group

What Is a Content Component?

A *content component* is the smallest unit of content that can be mixed and matched with other components to create an output. For example, rather than creating an entire chapter of a book by starting at page 1 and writing to page 30, content creators write, store, and manage the content in small chunks. Then, at the point of delivery, the chunks are put together in a certain order to create the final output.

The component is created, updated, managed, and retired from one—and only one—place. Only one version of every piece of content exists, and that version is used every place it is needed.

---

[6] "Rethinking Retail: Insights from consumers and retailers into an omni-channel shopping experience" (Infosys 2013)

[7] "Personalization Development Study" (Monetate 2017)

By creating small, reusable content components and tagging them consistently and correctly, you can make sure you deliver only the right pieces of content to the right person at the right time.

What Is Content Reuse?

Content reuse is the practice of creating one content component and then using that component everywhere you need it.

This single component is created once, approved once, stored once, translated into each language once, and retired once. If the information changes, you update the component in one place. Every experience that reuses that content component is then updated automatically.

This single piece of content is used over and over again to create many outputs and experiences. The component may be reused manually by a content creator who includes the content in a larger document or experience. Or it may be reused automatically by a system that matches the content to a customer.

If you don't have reuse, you can't personalize, because you are simply creating multiple instances of the information. We certainly can't do that at scale.

—Ann Rockley, CEO, The Rockley Group

How Reusing Components Enables Scale

Reusing components greatly reduces the cost and time required to create and manage content. Content component reuse also provides a number of benefits that enable you to scale content operations and delivery. Even before companies started delivering personalized experiences, they looked to reuse to achieve a number of operational goals:

- Create, review, approve, and translate content only once
- Assemble deliverables quickly
- Automate multichannel and omnichannel publishing
- Eliminate duplicated effort
- Reduce desktop-publishing time
- Say the same thing the same way, everywhere you need to say it
- Prevent content drift and copy/paste errors

By achieving these goals, companies can reduce time to market for both content and the products that content supports. They can also improve the quality of their content, both in the source language and in localized versions.

The Problems with Personalization

Companies have used a variety of methods to try to provide personalized experiences. Marketing uses personas and journey maps. Technical communications and training use conditions and variables. Just about everyone has thrown a new tool or two at the problem. But successful personalized content at scale remains elusive.

Personas Don't Work

Personas were first introduced as a way to narrow down the number of unique outputs we need to create to personalize content. They are almost as old as digital marketing itself.

A persona is a representation of a stereotypical customer. Personas have become popular as a way to give marketers a person to write for. Unfortunately, the result is content that is created for fictional stereotypes rather than content that provides the right information at the right time for a real-life customer.

Companies often go to great lengths when developing personas, spending lots of time and money creating them. Most companies create these archetypes by using collected data and web analytics, imbuing personas with multiple qualities:

- Name
- Age
- Gender
- Children
- Hobbies
- Income level
- A need that aligns with one or more products or services

Sometimes, personas specify details such as race, marital status, sexual orientation, parenting status, and more. They often include a photograph or image of the fictional consumer.

After personas are created, targeted content is developed for each. The hope is that information based on these stereotypes can be used to personalize the user experience.

Personas can help direct and limit content when you don't have enough information about a person to provide a unique experience. But as most organizations are finding, writing to a persona falls short of providing a truly personalized experience. Such an experience is unique to the person reading your content—not to a stereotype that they might fit into.

Here are a few examples of personas for a fictional video camera manufacturer.

Eddie is a 70-year-old male. He is married. He recently retired and wants to turn his videography hobby into a part-time business. Eddie has spent modest sums on consumer-grade electronics over the years, but now he's ready to invest in pro-level equipment. He's budgeting $15,000 to start and $7,500 per year for electronics.

Jonathan is a 36-year-old male. He is married and has two children. He is interested in purchasing a video camera to capture special moments as his children grow. On average, Jonathan spends $3,000 per year purchasing electronics.

Pidge is a 25-year-old female. She is single. Her hobbies include hiking and photography. Pidge is interested in purchasing a video camera so she can capture her experiences hiking along mountainous trails. She is also interested in post-production software so that she can edit her videos. On average, Pidge spends $5,000 per year purchasing electronics.

Customer Journeys Don't Work

> We can create personalized content that doesn't know the boundaries of prescribed journeys. Smart people design content journey maps every day with good intentions. They create impressive and orderly graphs that say, "In this phase, this is the information they need. In that phase, this is the kind of content they need." But that's not how it works in the real world. Real content journeys look more like a bowl of spaghetti.
>
> —Mike Iantosca, Enterprise Content Strategist

Another way of creating semi-personalized content is to focus on each part of the customer buying journey. Instead of targeting a persona, content created for a journey targets where a person is in their process as a customer.

The customer journey has various stages, usually specific to your product, service, or brand. But all customer journeys have one thing in common: At each point of the journey, the customer needs unique content.

For example, a customer sales journey might contain these stages:

- Awareness
- Analysis
- Decision
- Retention
- Championing

A customer post-sales journey might look like this:

- Install and configure product
- Verify installation and configuration
- Troubleshoot and resolve issues

The required content changes at each stage of the journey, regardless of the type of journey the customer is on at a given point.

Sometimes, companies combine personas and journeys, creating a different journey (and different content) for different personas at different parts of each persona's journey. You can see how the amount of content needed can grow quickly and exponentially.

Tools Alone Don't Work

Many companies have invested in new systems that promise to deliver personalized experiences at scale. Then they are shocked when they don't immediately get the results they expect.

What these companies seldom realize is that if you do not change how you think about content, tools can do only so much. For example, you can't show a personalized piece of content to a customer if you don't use components to write content. You can't optimize content for personalization if you don't standardize your terminology. You can't match content to the customer if you don't tag the content appropriately.

When it comes to delivering personalized experiences, changing your content is as important as—if not more important than—deploying new tools.

 This is not a technological change. This is a cultural change.
—Charles Cooper, Vice President, The Rockley Group

Old Paradigms Don't Work

As you can see, the problem is that companies focus on the wrong things:

- Content delivery instead of content creation and management
- New tools instead of new ways to create content
- Content silos instead of a holistic process

Starting at the End, Not the Beginning

Time and again, we see companies focus on the end of the content process: outputs and delivery. Too many content projects begin with the end. This approach limits the content as well as the way you create and manage it. Companies that start at the end must try to apply personalization to content that already exists or that was created using old methods and mindsets. Such content is rarely reusable. And reusability is the key factor in scalability.

To succeed with personalization at scale, you need to focus on the creation and management of the content, free of the limitations of how that content will be delivered. Focusing on content in a format- and output-free way enables you to create small chunks of important, reusable content.

New Tools, Old Content

As humans, we tend to be attracted to the newest bright, shiny object. The same goes for the latest technology and tools to hit the market. The second content-based reason that personalization fails is the uncanny lure of new tools.

Companies invest in new tools, thinking that technology will solve the problem. But new tools plus old content simply does not work. As Val often says, "If you throw expensive tools at crappy content, you end up with expensive crappy content."

No tool on the market is going to take your inadequate content and make it spectacularly reusable. It just isn't going to happen. This isn't to say you won't need new tools; you might. But before you invest in them, invest in fixing your content.

Same Silos, New Paradigm

Content creators have been talking about silos for decades—longer than either Val or Regina have been in the industry. If you want to successfully personalize the customer experience at scale, you *must* work across silos. The customer does not care which group in your company is responsible for creating which content.

Content across your company needs to flow seamlessly and look like it comes from one company. Otherwise:

- If different content components use different terminology, grammar, and aesthetic, they cannot work together seamlessly when mixed and matched.
- If your marketing people have worked hard to develop a company voice but the product documentation team doesn't use it, then your customers experience friction when they move between marketing content and documentation.
- If documentation, training, and support use different words to mean the same thing, customers may become confused or even angry at the discrepancies.

The catch is that only by making content entirely consistent can you reuse it to provide a frictionless, seamless personalized experience.

Summary

Herein lies the personalization paradox:

To create nimble, reusable content components that can be combined, on the fly, in different ways for different people and different devices, *you must standardize everything about your content.*

CHAPTER 2
The Big (Data) Picture

This book focuses on what you need to do and understand to create personalized content at scale. We now know that to personalize content at the point of delivery, you need to begin with small, nimble, standardized content components that can be mixed and matched on demand. Standardizing the inputs enables you to customize the outputs.

But how do you figure out which content to deliver to each person?

There are three basic ways that systems assemble and deliver a personalized set of components:

- The system already has information about the content consumer.
- The system asks the consumer for information.
- The system monitors the consumer's online identity to figure out what the consumer needs.

Personalization is now so expected that most people are willing to share personal data to get a personalized experience. According to a Salesforce study, 57 percent of consumers said they would share personal information to get a personalized offer. And 62 percent said they are okay receiving personalized offers from companies based on things that they've purchased in the past.[1] It does look like we are loosening our grip on what we share as we increase our expectations of receiving personalized experiences.

Different industries have different rules—and different customer expectations—around personalization. All industries need to find the right balance between tracking enough data to provide better content experiences and tracking too much data. At best, over-tracking can bury your team in data and scare off your customers. At worst, you can get in big trouble with regulatory agencies, governments, and other institutions around the world.

---

[1] "It's Personal, and It's Business: Using Retail Personalization to Connect with Customers" (Salesforce)

You Know What They Want

In many instances, a company already has collected a significant amount of information about the content consumer, particularly when it comes to large purchases. Take automobile manufacturers. When you buy a car, the vendor collects an abundance of information:

- Name
- Address
- Phone number
- Car details:
  - VIN, which includes year, make, model, color, features
  - Warranties
  - Add-ons (e.g., floor mats, trim upgrades)
- Trade-in car (if you have one):
  - VIN, which includes year, make, model, color, features
  - Mileage
  - Condition
  - CarFax report of accidents, maintenance, history
- Banking information (if you apply for financing)

Using this plethora of information, the automaker can assign a variety of metadata to you and match that metadata to the content components you might need in the future.

Here's an example: Let's say that two years ago, you purchased a brand new Acme Sedan. Now, you need to replace the battery in your key fob. Thanks to the information collected at the time of sale, Acme already knows what model sedan you purchased and the key fob that came with your car.

When you log in to the Acme website or support portal to find instructions for replacing the key fob battery, the system immediately identifies who you are and what you purchased. When you search "replace battery in key fob," the system provides you with exactly the instructions you need for your year, make, and model of car.

You shouldn't need to provide any additional information. You shouldn't need to search among different style key fobs or select your year, make, and model. You shouldn't need to dig through the entire operator manual just to find the right instructions. With the correct inputs and metadata, the automaker should be able to deliver an extremely fast transaction—and you're on your way.

> If information is properly chunked, it can go to that person. The system knows who they are. The system knows their skill level. The system knows all sorts of stuff about them. If that service information is properly segmented, they can get the information that they need right there.
>
> —Charles Cooper, Vice President, The Rockley Group

Asking for Information

Being asked for information is ubiquitous in our digital age. Providing information to narrow a query is an everyday occurrence. Most of our online support interactions start by answering some questions for a chatbot so that it can locate the information we need.

But how many times have you interacted with a chatbot, only to be presented with either too much or the wrong information? Unfortunately, this happens all the time.

Two things can cause such a situation:

- The company hasn't chopped content into small, individually accessible components.
- The company hasn't created a robust taxonomy and metadata strategy to enable its system to pinpoint and serve exactly the content components you need.

Either way, you are left with a less-than-expedient experience—and too much work ahead of you.

Let's return to our Acme car key fob example. This time, suppose you purchased the car used from an individual, so the manufacturer doesn't know anything about you. When you venture to the Acme support site, you are met with a chatbot.

The conversation probably goes something like this:

- **Chatbot:** Hello! How can I help you today?
- **You:** I need to change the battery in my key fob.
- **Chatbot:** I understand you need to replace the battery in your key fob. I'd be happy to help you with that. But first, I need to ask some questions.
- **Chatbot:** What is the year, make, and model of your car?
- **You:** 2018 Acme Sedan
- **Chatbot:** You said you have a 2018 Acme Sedan. Is that correct?
- **You:** Yes [*You are starting to get annoyed by now.*]
- **Chatbot:** One moment.
- → *time goes by* ←
- **Chatbot:** Here is the 2018 Acme Sedan Owner's Manual.

How many of us have experienced this over and over again? Rather than personalizing the answer and providing only what you need, they give you the entire owner's manual and make you search.

Always Watching?

Companies use a variety of techniques to learn about us and make decisions about what information to show us. The creepiest way companies get information about us is by stalking our online habits. Targeted advertising in social media is an extremely successful example of the use of this technique.

Companies collect information frequently, using many vehicles. Here are just a few:

- Virtual assistants
  - Siri
  - Alexa
  - Google Assistant
  - AliGenie
- Social media
- Browsing activity
- Interacting with smart devices (via the Internet of Things, or IoT)

Admittedly, when a company uses our habits to determine how to personalize our content, it makes a somewhat educated guess about what we care about. They aren't always successful. We've all experienced social media platforms showing us completely irrelevant ads based on a one-time Google search.

 I think that's one of the biggest challenges: balancing people's right to privacy and people's conditioned demand for exceptional experiences and having things delivered in a personal way.

—Harvey Turner, COO, Pilot44

The Impact of GDPR on Personalization

The General Data Protection Regulation—GDPR for short—is a European Union (EU) law that was implemented in May 2018. GDPR applies to any data storage that contains a person's name or other identifier. Even though GDPR is an EU regulation, many companies in other parts of the world comply with it.

The purpose of GDPR is to provide guidance to companies that collect personal information and to give individuals more protection and rights over how their personal information is used. GDPR governs how personal information can be collected, stored, and used. The regulation states that personal data should be secure and accurate. Companies must explain how they will use such data and must restrict their use to the minimum needed to do the specified job. Most importantly, companies must get consent from each individual before collecting and using their personal data.

GDPR contains seven principles for the collection, storage, use, and destruction of personal data:

1. Lawfulness, fairness, and transparency
2. Limitation of purpose
3. Minimization of data
4. Accuracy
5. Limitation of storage
6. Integrity and confidentiality
7. Accountability

In addition, individuals can use a subject access request (SAR) to ask a company for the information that has been collected about them, how it is being used, and that it be deleted. Companies must respond to a SAR within thirty days of receiving the request.

Clearly, GDPR affects our ability to collect information about content consumers. Here are some issues to consider when collecting personal information under GDPR:

- A visitor to a website must explicitly agree to the collection of their information if a company uses cookie tracking. This is probably the most obvious change since the inception of GDPR. Almost all companies must now post an opt-in bar at the top or bottom of a website.

- Consent forms must be easy to understand, provide links to the terms of service, and make opting out easy. You must also provide a consent form each time you ask for someone's personal information.
- Companies must disclose what they are doing with a person's information.
- Companies are permitted to collect only the information they need about a person.

It's not just the EU that has started to regulate what data can be collected and how it can be used. While we were writing this book, California enacted the California Privacy Rights and Enforcement Act (CPRA) of 2020. This act expanded the 2018 California Consumer Privacy Act (CCPA),[2] which expanded privacy protection following the 1972 amendment to the California constitution that defines privacy as an inalienable human right. The CCPA grants consumers several rights around the personal information collected during their digital journeys, including the right to know whether their information is being sold and the right to access collected information.[3]

Companies with customers in California must disclose the types of information they collect and provide consumers with a way to request their personal information. They must enable consumers to request the deletion of their personal information and must have a policy in place to respond to such requests. Companies must also provide a "do not collect my information" alert on their websites. Note that this law applies when the customer is located in California, regardless of where the company is located.

---

[2] https://www.jdsupra.com/legalnews/california-privacy-rights-and-65727

[3] "Analysis of the California Consumer Privacy Act" (Katz 2018)

By now we've probably all experienced the effects on consumers of both GDPR and the CCPA. Surfing the web, using social media, and even using mobile apps now involves more and more prompts to manage our personal information or to allow cookies.

But the news isn't all bad for companies seeking to deliver personalized experiences. Collecting information might be more difficult, but it is not impossible. There is something to be said for interacting only with people who specifically want to interact with you, as well as for complete transparency. After all, trust is one of the most important assets of any brand.

By asking for permission and disclosing your intentions, you can build a customer base that *wants* to be sold to via a personalized experience. You can increase your brand's value, trust, sales, and repeat sales.

Laws around privacy change frequently. In addition, such laws vary dramatically from country to country and region to region. Always consult an attorney to make sure your practices comply with the current laws.

> I personally don't care if you know my name or not, as long as you've anticipated my needs. However, there are definitely cohorts who would rather you didn't know their name or much about them at all. Yet, they typically respond to a curated experience as well as any other cohort. It reminds me of my dog when she was a puppy. She loved to fetch the ball and run it back, but then when you reach to take it, she's like, "No, don't take it—but throw it again." Like, "I want my privacy, but I also want you to anticipate my needs."
>
> —Marcus Hearne, SVP, Product & Solutions Management

Summary

For content personalization to work, you must correctly anticipate the content that your customer needs, at the time they need it, on the device they are using. There are three ways to do that:

1. Use information your company already has about the customer
2. Ask the customer for information
3. Monitor the customer's online identity to figure out what they need

CHAPTER 3
Artificial Intelligence and Personalization

 The other thing that is significant is the emergence of machine learning and AI to generate tailored, personalized, automated experiences. Ironically, to create a very personal and individual experience, unique and potentially different, you need incredibly well-structured content for delivery across all channels. Otherwise, you can't create the necessary rules to deliver an experience that is personalized from a business standpoint and feels very personal to your customers.

—Harvey Turner, COO, Pilot44

Today, most content deliverables are constrained by a predefined output type. Is the end product a knowledge-base article? An online help topic? An eBook? Output type is the first thing to standardize when you start creating personalized experiences at scale.

But the future of personalization will no doubt include artificial intelligence (AI) at a much more sophisticated level than we have today. AI has the ability to create both wide and deep cross-matrices of information, including weighted assumptions to increase the accuracy of personalized information. We believe that AI will eventually eliminate the need to create a standard output type. Someday, AI engines will be able to package the information the customer needs in whatever format makes the most sense.

AI systems excel at two things:

- Working with enormous quantities of data (or content)
- Processing data extremely quickly

In 2018, researchers at the University of California, Los Angeles unveiled a neural network that could solve complex mathematical computations at light speed.[1]

---

[1] "This AI Calculates at the Speed of Light" (Engelking 2018)

In addition to massive data and processing power, AI engines contain something called *machine learning* (ML). ML is the capacity of an AI engine to learn as it goes. In other words, once initially trained, AI engines use algorithms and statistics to keep learning, without needing additional input from people.

AI Applications

AI has many applications. Here are a few areas in which AI is deployed today:

- **Searching and planning.** AI engines can search through possible actions and react; for example, while playing chess.
- **Reasoning and knowledge representations.** AI engines store information in a representative way that can be reasoned about, and then used to make a conclusion. For example, the IBM Watson AI won on the game show Jeopardy.
- **Perception.** AI engines can interpret vision, hearing, and touch to help create systems that can act like humans.
- **Motion and manipulation (actuators).** AI engines can move and manipulate objects. This function is a sub-component of robotics.
- **Natural language processing (NLP).** AI engines can understand and act on parts of speech using techniques such as:
  - Language modeling
  - Summarization
  - Question answering
  - Dialogue systems
  - Parsing
  - Tagging
  - Speech recognition
  - Natural language generation (NLG)
  - Machine translation (MT)

Marketers have started using AI extensively to help personalize the customer experience. According to the fifth edition of the Salesforce State of Marketing report, marketers are using AI to:[2]

- Provide real-time next best offers
- Predict and personalize the customer journey
- Improve customer segmentation
- Automate social interactions

Any modern marketer who hopes to personalize content at scale would be well-advised to keep an eye on AI developments. In this chapter, we describe how AI works from a 50,000-foot level.

How AI Works

NLP is at the heart of most AI systems. NLP provides the AI system an understanding of the meaning and intent of content. And NLP chunks a sentence into its core components. For example, in the sentence …

The `.pdf` *file from Massachusetts may have information about the temperature in Fahrenheit.*

… NLP recognizes the following information:

- `.pdf` is a file extension.
- *Massachusetts* is a place.
- *may* is a modal verb.
- *temperature* is a noun.
- *Fahrenheit* is a unit of measure.

Warning: AI cannot fix crummy content

AI engines can do a lot of things. But they cannot fix poorly written content. AI can quickly locate and serve up poorly written content but cannot change or correct it in any way.

---

[2] *State of Marketing* (Salesforce Research 2019)

It is up to us humans to ensure that our content is readable, understandable, and factual. And, as we discuss throughout this book, the more you standardize your content—from the words you use to the way you use them—the better the quality of your content will be.

Setting up an AI System

Setting up an AI system involves four stages:

- Upload
- Curate
- Ingest
- Train

Upload

The first thing that an AI system needs is a corpus of data. If you deploy an AI engine at your company, you will likely upload all the content that you have to the engine. And when we say *all*, we mean *all*:

- Engineering specs
- Manufacturing requirements documents
- Test documents
- Technical documentation
- Training documentation
- Knowledge-base articles
- Sales enablement material
- Marketing content
- HR communications

… and just about anything else you can get your hands on. The more information you upload to the engine, the better job you can do of training the engine. And the more you can train the engine, the better job it will do for you.

Curate

At some point, either before or after you upload the corpus, someone (or a team of someones) needs to curate the content.

During the curation process, subject matter experts eliminate content that is outdated, inaccurate, or not good for whatever reason. You don't want to train your AI engine with incorrect or old information.

You also want to eliminate bias in your AI training content. Unfortunately, prejudicial bias has crept into almost every internal and external system we have today. We use terms and phrases that are inherently biased, often without even realizing it. In turn, those terms and phrases show up in search results and in the content itself.

Here are some examples of prejudicial terms that are quite common:

- Man hours/weeks/years
- Master/slave
- Chairman/chairwoman
- Blacklist/whitelist

Imagery in your corpus can also train the AI engine to be inherently biased. For example, if your corpus of photos contains mostly male firefighters, the AI engine might presuppose that all firefighters are male. If your corpus of photos contains mostly female teachers, the AI engine might presuppose that all teachers are female. And so on.

Curation is a big job and one of the most important steps to get an AI engine up and running properly.

Ingest

Ingest is an interesting term; it sounds quite biological. After your corpus is uploaded and curated, the AI engine goes through an initial process called *ingestion*. During ingestion, the engine sorts through the corpus:

- Organizing information
- Categorizing information
- Semantically tagging information so that it can be retrieved later

There is no human interaction with the system while it ingests the corpus. The AI system acts on its own and creates its own organization and tagging system. This information is internal to the AI system and unavailable to us humans.

Train

After the corpus has been ingested, the actual training of the AI system takes place. During the training process, subject matter experts use pairs of questions and answers to teach the engine how to process connections between pieces of information.

These question-and-answer pairs are called the *ground truth*. In essence, when you train an AI engine, you teach it about the linguistics of a certain domain of knowledge. For example, we use different words and sentences to talk about travel than we do about oncology. Using ground truth questions and answers, the AI engine learns all about the linguistics of the domain, understanding the meaning and intent of the sentences and questions.

After the engine is trained using the ground truth, it can then use the entire corpus for its purpose. For example, an AI engine is frequently used to look up information, much like a web search. The difference is that a web search returns thousands and thousands of hits that may or may not answer the query. A well-trained AI engine with a carefully curated corpus is much more likely to return the correct information the first time.

Over time, using ML, the engine continues to learn more and more about the linguistic domain. During the ML process (which is continual), the AI system continues to make connections without specific human interaction.

 With a corpus of intelligent content that consists of metadata-rich, containerized components and micro-components, you can quickly train AI auto-classifiers to assign taxonomic values to those containers and automatically generate ontological models at scale that are far more accurate and precise than just leaving it up to the AI engine to randomly generate from unstructured big data. You can then iteratively refine them instead of hand-crafting one-offs that don't scale. What we are really after is automated assembly and delivery of the precise combination of content components that are tailored to the user's intent and task at hand. We're not simply looking to create yet another narrowed search experience that is only an incremental improvement on a failure-mode content model. That doesn't move the needle.

—Mike Iantosca, Enterprise Content Strategist

Processing Information

AI engines use an artificial neural network (ANN) to process requests and provide results. ANNs work a bit like the neural network in the human brain.

Consider the following example—a true story.

Val recently needed to travel from the San Francisco Bay Area to New York City. When selecting her flight, she had many criteria to consider:

- Did she want to fly out of the San Francisco, Oakland, or San Jose airport?
- Did she want to fly to John F. Kennedy, LaGuardia, or Newark airport?
- Did she want to fly non-stop or were stops okay?
- Did she want to fly on a red-eye flight or leave California early in the day?
- What time did she want to fly home?
- Were her plans flexible enough to accommodate better prices on different days?
- Were there any airlines that she wouldn't fly, no matter what?
- Did she want to travel using frequent flier miles or pay for the tickets?
- Did she want to travel on an airline on which her frequent flier miles accumulate?
- Did she want to travel business class or economy?
- How much was she willing to pay?

As Val went through this list, she assigned weights to each criterion and biases to each connection between criteria. For example, flying non-stop was more important to Val than the airport out of which she flew. She preferred flying into John F. Kennedy over flying into Newark. Using her frequent flier miles was important—*if* they enabled her to travel business class. She definitely didn't want to fly on a red-eye unless it was the only choice.

Making a decision by combing through all the information, the connections between criteria, and the weights of each criterion is a complicated process. But it is a process that we all use naturally when we make a complex decision.

In a similar way, an ANN assigns a weight to each node (or criterion) of information and a bias to each connection between nodes. Using a complex web of nodes and connections, called *hidden layers*, and an algorithm for decision making, the ANN can produce one or more results, along with a likelihood that a given result is the best choice.

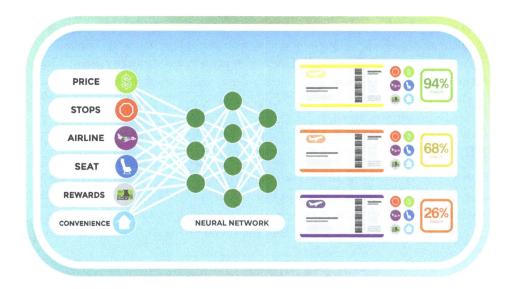

Using an AI Engine to Personalize Content

If an AI engine knows Val's preferences and understands her weights and biases, it can automatically produce the perfect airline flights for her, all the time. As the system learns more about what she likes and does not like (based on selections over time), it can factor in even more information to produce a personalized itinerary just for Val.

For example, she might have more difficulty using frequent flier miles in summer because more people travel, so fewer frequent flier seats are available. Or perhaps as she uses frequent flier miles, she has fewer of them to spend. Therefore, the AI system needs to adjust the weights and biases to accommodate these changes.

Because AI is always learning, it can adjust the way it acts on information as time goes on. If Val starts flying exclusively on one airline, the AI system can start prioritizing that airline. If she starts flying only business class, regardless of airline, the system learns this, too, and adjusts. For Val, getting only flight information that meets her criteria—in other words, a personalized experience—would save countless hours and searches.

Many companies all over the world are deploying AI systems to provide a level of information that was previously unavailable, trapped in far too much content for any human to parse. As more and more AI and ML engines come online, more of our preferences will be noted and tracked. And personalized content will become more accurate and easier and faster to deliver. AI helping with personalization is not an *if* but a *when*—and we think that time is likely sooner than later.

Summary

The future of personalization will include artificial intelligence. AI can create wide and deep cross-matrices of information, including weighted assumptions to increase the accuracy of personalized information. It's already in use for searching, planning, reasoning, perception, object manipulation, and natural language processing. Standardized content is important in training AI systems.

CHAPTER 4
What Is Content Standardization?

Content standardization is the practice of creating content that follows a shared set of rules and guidelines. It is how you personalize content at scale without confusing your customers. Standardized content includes rules for every aspect:

- Words that are preferred and words that are not allowed
- Style and grammar rules
- Voice, tone, and branding guidelines
- How sentences combine to create paragraphs
- How paragraphs combine to create sections or content components
- How components combine to create the final output

Creative people often have a strong reaction to the word *standardization*. The term connotes a realm of grey, rectangular, boring content. It triggers feelings of confinement and resistance.

On the one hand, it's true that standardization affects creativity. Creating content that adheres to established standards calls for a little less creativity. But business writing is not about creating poetry or inventing an entirely new way to quantify the universe. **Business writing is about producing effective content that serves the company's needs and can provide a personalized experience for customers, at scale.**

On the other hand, standards free content creators from repetitive decision making, as well as the need to redesign the consumer experience from the ground up, every single time. Standards provide the measurements, blueprints, and principles that enable creative people to build personalized content experiences much more quickly and easily.

A simplified content strategy is an asset for the company.
—Anna Schlegel, Strategic Globalization and Content Strategy Leader, Author, and Speaker

What Are Standards?

Standards are the rules your company agrees to follow for all published content. Standards must be documented and adhered to. They define how content is written, stored, managed, published, and retired. Some standards are strict; some are loose. Every company has its own set of standards that content creators are supposed to follow.

When you clearly document and enforce standards, you are rewarded with:

- Consistent content
- One company-wide voice
- Brand cohesion
- Improved readability
- Reduced translation costs
- Increased translation quality
- Decreased time-to-market

An Analogy: Building an Experience to Make Customers Feel at Home

In many ways, creating personalized content experiences is like creating personalized dwellings. Generally speaking, there are three types of new homes in the United States, each with varying levels of standardization:

- **Multi-tenant dwellings,** with multiple apartments that are identical in terms of fixtures, appliances, flooring, and so on. Apartments are the most standardized, and least personalized, of all housing types.
- **Custom homes,** in which everything is personalized and unique. Custom homes are the least standardized, but the most personalized, type of home.
- **Planned housing communities,** which comprise a handful of floor plans that buyers can personalize to an extent by selecting fixtures and finishes from a standard set of options. These types of communities allow the housing industry to create personalized homes, at scale.

Multi-tenant Dwellings: High Scalability, Low Personalization

The most standardized, least customized type of US housing is the multi-tenant dwelling (or apartment building). Most of these buildings are created with no particular tenant in mind.

When a builder builds an apartment building, there is usually no personalization of individual units. Most of the time, the occupants of each apartment are not known in advance. The builder buys everything in bulk. Every apartment has the same:

- Windows
- Doors
- Bathroom fixtures
- Kitchen fixtures
- Appliances
- Flooring
- Wall color
- Doorknobs

… and so on. This method saves a lot of money and time.

Figure 4.1 – Four apartment interior layouts (source: Apartments.com[1])

---

[1] https://www.apartments.com/edgewood-court-chicopee-ma/yw6pzzr/

In most apartment buildings, the units that stack up vertically each other have the same layout, number of bedrooms, and so on. Across any given floor, the floor plan of each unit might be different. In both cases, the finishes are identical (see Figure 4.2 and Figure 4.3).

Figure 4.2 – Modular floor plan 1

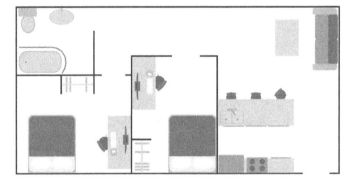

Figure 4.3 – Modular floor plan 2

The standard apartment in a large multi-unit building is one size fits all. Likewise, without any personalization, we end up delivering the same content to every consumer, without taking their individual needs into account. Serving everyone the same, albeit standardized, content is relatively inexpensive and easy to do, but it fails to provide any type of personalized experience.

Custom Homes: High Personalization, No Scalability

On the other end of the spectrum are custom homes. Everything in a custom dwelling is individually selected: doors, doorknobs, moldings, windows, doorstops ... the list goes on. Everything that goes into the home is unique, purchased specifically for the individual that will live in the home.

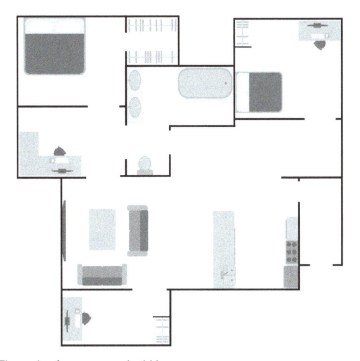

Figure 4.4 – Floor plan for a custom-build home

Custom homes are not always luxurious mansions. Regina can attest to this fact: She lives in a 1,000-square-foot custom cottage that someone built as a do-it-yourself project in the 1940s or 1950s. Nothing is standard, right down to the sewer pipes. A simple toilet repair can easily turn into an excavation of 15 feet of pipe. The electrical system is wired so oddly that some wires go nowhere. Small wooden planks nailed to burlap-on-plaster walls prevent any plugs with an AC adapter from fitting into the power outlets. Every window frame is a different size, necessitating custom fittings whenever a window screen rips. And good luck buying a window fan.

As a result, Regina cannot repair, update, or modernize this house within any kind of reasonable budget or time frame. Nor can she replicate the house at scale.

This is the dark side of personalization. Only Regina and the original builder can possibly love this house—nobody else will put up with it! You simply can't force a personalized experience onto the wrong person.

A custom home is the opposite of a standard apartment. Everything is unique … and typically more expensive and difficult to design, build, and maintain. Many companies try to create a completely personalized experience for each person. Although doing so is possible, it is time consuming, expensive—and impossible to scale.

Planned Housing Communities: The Best of Both Worlds

There is a third option: the planned housing community. In a planned housing community, a set of options for various components of the house offers a measured level of variety. For example, homeowners can choose their favorite from:

- Six floor plans
- Three styles of kitchen cabinet
- Four types of countertop
- Two types of doorknob
- A defined selection of hardwood, tile, or carpet

… and so on.

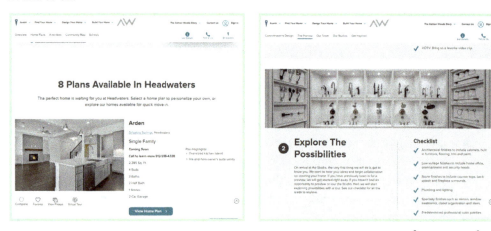

Figure 4.5 – Planned housing community (source: ashtonwoods.com leftimage[2] rightimage[3])

Building houses in this way provides a balance between a completely standardized home and a completely customized home. It provides various options for personalization while limiting the

---

[2] https://www.ashtonwoods.com/austin/headwaters
[3] https://www.ashtonwoods.com/design-studio/

options to a reasonable number of choices. The builder benefits from quantity of scale, and the occupants benefit from details that personalize each home.

A planned housing community is the best model for creating personalized content experiences. When you build an output—be it web page, online help system, knowledge base, or training course—you simply select content from a specified set of standardized components. You do not need to create a new output for each and every person. However, you still have the opportunity to provide a customized, personalized experience that matches what each person needs. And you can do so at scale.

> Val lives in the wine country in Northern California. In 2017, a firestorm called the Tubbs Fire ravaged the area all around her house. Several neighborhoods were incinerated. Miraculously, Val's home did not burn.
>
> When rebuilding began in 2018, Val noticed an interesting trend. One neighborhood previously had tract houses, all of which were gone. A developer purchased about 20 lots in the tract and built semi-custom homes, just like a planned housing community. There are only a couple of different floor plans, but the homes look unique because of the finishes. These homes were built quickly and were fully occupied within the year.
>
> Further down the road is an area of custom homes. Each home is being rebuilt by a different contracting firm. Each home is completely unique. Three years later, most of the custom homes are still under construction. It takes a very long time, and a lot of money, to build a completely personalized, custom home.

> You have these standardized rules. But in those rules you can customize to your heart's content. You can pick a big rain showerhead with 300 little openings or a teensy handheld showerhead. The connector's still going to be a three-quarter inch, and it's going to connect to every shower. And that's what good standards do: They allow for interoperability. The foundational parts are the most basic things, like the terminology, the style, the grammar, but if you don't have that solid foundation, you're basically going to be assembling junk.
>
> —Mike Iantosca, Enterprise Content Strategist

Brand Standards

We often find that companies put significant effort into developing brand standards. These standards include design standards for how content looks when published: which fonts to use and where, primary and secondary color palettes, and logo sizes and placement. Brand standards guide marketing and digital marketing in creating a cohesive look and feel for customer-facing content.

But content silos may not be aware of corporate brand standards. We've worked with many technical communications departments whose user manuals, installation guides, API document-ation portals, and developer websites have nothing in common with corporate brand guidelines.

Consistent design affects the customer experience. We've all been to corporate websites that have a totally different look and feel from the company's product support site. The difference can be so strong that we aren't 100-percent certain we're accessing the right content—or even the right company.

Traditions Are Not Standards

Most of our customers have some standards in place before they reach out to us for help. They have a style guide. Or they have templates in place to control what documents look like. Or they have a term list, maintained by one or two industrious copy editors who find themselves correcting the same issues in the same type of content over and over again, because the writers don't check the list frequently enough (or at all).

In such cases, we often discover that content teams have been relying on departmental tradition and personal preference rather than approved standards. Each silo (i.e., department) and even each team within each silo has its own way of doing things. Differences in tools, processes, and individual preferences all infuse the content with inconsistencies.

Other times, we find that existing standards aren't documented. Instead, they are contained in the smart brains and creative imaginations of various people in the company. Such standards are treated like company lore—an oral tradition passed down from person to person.

Lack of approved, documented (and enforced) standards can negatively affect your content—and the customer experience—in many ways:

- Mismatched assets
- Decreased quality
- Brand degradation
- Higher translation costs
- Longer time-to-market
- Increased confusion

Worst of all, non-standardized content that does not conform to your brand can keep customers from reaching the most important goal of customer interaction: becoming an advocate for your brand.

Think about your content teams. How much does your company rely on undocumented knowledge for writing standards, terminology, process, and workflow?

The biggest problem with standards maintained via such knowledge is that they don't scale. Not to mention, they can be unreliable and rarely result in consistency within a silo, much less across silos throughout the company.

We've had a number of authors tell us over the years, "We do it this way because that's how we've always done it." They know that their methods do not help them deliver, much less personalize, content effectively or efficiently at scale. They just don't know what to do instead. They don't know what or how to change.

Separate Format from Content

To provide a cohesive, personalized experience at scale, the output needs a standard look and feel. How can you achieve this goal when reusing small components, mixing and matching them on the fly?

The solution is to separate *format* from *content*. When we create a content component, we do not know where it will be used or reused. The same component might be used in output for a web page and output for an online help system. By separating the format from the content, we are able to focus only on the content.

The format is applied to the content during the publishing process. This enables the same content component to look different on different devices or in different outputs. Separating content from format is an intrinsic part of delivering personalized experiences at scale.

What Needs to be Standardized?

Standardizing content to create a personalized experience might seem counterintuitive at first. After all, when we think of a personalized experience, we think of unique content that is created for and delivered to a unique individual. But as our housing example has illustrated, creating unique content this way simply cannot scale.

To scale, you must mix and match pieces of content that are reused in different ways for different people. And the only way for this mixing and matching to work is to ensure that all the dimensions of your content are standardized.

To create a personalized experience from a collection of standard components, those components need to be standardized across five dimensions:

- Output type
- Word
- Sentence
- Paragraph
- Component

Output Type

Outputs are assemblies of content that get published and delivered to customers. A standardized output type provides a framework for each experience you want your customer to have. An output type is like the floor plan of your house. You need to know the number of bedrooms and bathrooms you want before you can select the floor plan.

Word

Words are the smallest standardized content unit. A collection of words is called *terminology*. Standardize terminology so every piece of content you create uses the same word to mean the same thing. Terminology is like your fixtures—it's the smallest detail that makes the house a home.

Sentence

The standards that govern plumbing and electrical systems define how various fixtures are combined. Grammar and style rules govern how words are combined into sentence. For the system to work, everyone needs to follow the rules.

Paragraph

The paragraph is where you define the tone and voice of your company. A consistent aesthetic builds your brand and gives your home a distinct personality.

Component

The component is an independent unit of content that can be combined with other components to deliver an experience. A component is a room in your house.

The next chapters focus on how to standardize each of these dimensions.

Summary

Standardization is what makes personalization scalable. To standardize content at scale, you mix and match small components of content at the point of delivery. For that content to make sense, all the words, sentences, paragraphs, and components must follow the same rules. Otherwise, you risk confusing your customer.

There are five dimensions to standardize:

- Output types
- Words
- Sentences
- Paragraphs
- Components

Standardizing these dimensions is key to successful personalization.

CHAPTER 5

Managing the Experience: Standardizing Output Types

Output types are the assemblies of content that you publish for delivery to customers. Output-type standards define:

- The type of content to include in an experience
- The order in which to include the content
- Which content is required and which content is optional
- Which content to reuse every time
- Which content to create new every time

A standardized output type provides a framework for each experience you want your customers to have. It's the invisible structure that creates consistency and usability for the customer.

Thanks in part to personalization giants like Amazon and Google, it's common to envision a personalized experience as being a completely customized journey, with every piece of content matched exactly for an individual and delivered at exactly the point that they need it. After all, isn't that the whole point of personalization?

Yes and no. The whole point of personalization is to give customers the content they need, when they need it. Standardizing output types helps you meet that goal.

Output Types: The Floor Plan

Your content's *output type* is like your home's floor plan. Before picking flooring, cabinets, or faucets, the first thing a home buyer needs to decide is how many rooms they want in their house. A family with three adults, five children, and two dogs has very different needs than a family with one adult and a goldfish.

In a planned housing community, home buyers can choose from a variety of floor plans. Hopefully, at least one provides the right experience for the family. Every house with the same floor plan has the same types of rooms in the same location. You don't build a Model 123 with the kitchen in the front of the house and then a second Model 123 with the kitchen in the back of the house. Rather, the kitchen is in the same spot in every Model 123 plan.

This isn't to say that the home isn't personalized. One Model 123 might have pastel colors, stainless steel appliances, and window shutters. Another might have neutral colors, white appliances, and long drapes. This is part of the personalization process. The content may differ even though the organization of every Model 123 is the same.

What Is an Output Type?

An out type is any type of deliverable that you create with content. Table 5.1 shows examples of some common output types.

Table 5.1 – Common output types

Marketing	Documentation	Training	Support
• Company website • White paper • Case study • Email or series of emails • Product video • Social media campaign	• Online help system • Equipment manual • Installation guide • System administration guide • Quick start guide • Knowledge base	• eLearning course • Instructor-led presentation • Handout • Workbook • Webinar • Demonstration video	• Knowledge base • Frequently asked questions web page • Response (typically via email, web page, or both) • Customer service telephone script • Chatbot answer • How-to video

Let's look at two of these examples in greater detail:

1. An equipment manual delivered as a PDF
2. A product web page delivered as a dynamic web page

Equipment Manual

It's been the norm for decades that in order to create a new equipment manual, a content creator would duplicate an existing manual and then "make a few tweaks."

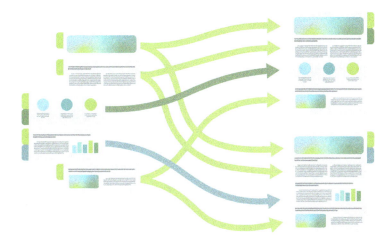

Even in a small company with only one or two technical writers and the best of intentions, differences very quickly creep in from one manual to the next. Some manuals start with an introduction and some don't. Sometimes the equipment manual includes setup instructions and sometimes that information is provided in a separate document. Some manuals include an appendix of useful reference information and others don't. Even things as simple as what information is included on the cover page can get out of whack.

These inconsistencies force customers to work harder to access the information. They look in the equipment manual for setup instructions and can't find them. They go online to search for instructions and discover that the setup instructions for this machine are in a separate document, which they now have to find. But for another machine from your company, the setup instructions are provided at the end of the equipment manual.

With a standardized output type, every equipment manual includes the same type of content, in the same order. All required content is always present. Any optional content that is included will also be included in the same place in each manual. Definitions, descriptions, warning statements, procedures, and other content that should be identical across multiple manuals is identical, because the standards show where to reuse content. Unnecessary content is excluded, because the standards make it clear for content creators not to include it.

Customers do not have to put in any extra work to find content from one equipment manual to the next. Customers do not experience mixed messages or have to figure out which set of almost-but-not-quite-the-same instructions to follow. Their experience may not be personalized in the sense of matching content to the customer at the moment the customer needs it, but it is optimized.

Product Web Page

A dynamic web page is a page that is assembled "just in time" when the customer visits.

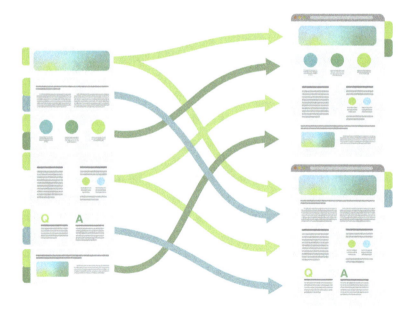

With a static web page, content is created for the page and every customer sees the same content. In contrast, a dynamic web page assembles several blocks of content at the moment the customer arrives. Different customers may see different blocks of content.

Regina worked with a hardware company that was piloting a handful of simple personalization options. One of the goals with their product web pages was to provide existing customers with one block of content and prospective customers (called "guests") with a different block of content. The rest of the page was the same for everyone.

Regina's team was asked to define the standards that would guide hundreds of dynamic product web pages. The content itself would be different for each product page, but the structure of the page would be the same across all products. If a guest became a customer and returned to the page in the future, the page would automatically show them the customer content.

The standardized output type of "product page" allowed the customer to scale what would have been impossible to deliver without standardization.

Why Standardize Output Types

In personalized content, the output type determines the content that forms your final deliverable and the order in which that content appears. It's how you define which pieces of content your personalized content delivery system can choose from when matching content to customer.

Standardizing output types also provides the key to knowing what types of experiences you can deliver. Today's publishing systems make it so easy to pull web pages together that many companies have reached a point where they have tens of thousands of dynamic links to manage. They're building houses quickly without first drawing a floor plan, and the result is even more difficult to manage than you're imagining right now.

The structural consistency provided by standardization makes an enormous difference to customers who need to find and use your content.

Standardized output types help you:

- Improve the customer experience
- Streamline content operations
- Automate publishing
- Prepare content for AI

Improve the Customer Experience

Standardized output types use consistent navigation, organization, and information types. Even in long, linear document formats such as PDF or print, customers learn very quickly where to find the information they need.

Consistency across a body of content improves findability and usability. This has been true since the beginning of written information. Consistent content improves the customer experience.

Not all output types adapt well to a personalized environment. For example, a corporate website can be highly personalized. A printed user guide, tucked into hundreds of thousands of boxes containing identical Bluetooth headsets, cannot.

Streamline Content Operations

Standardizing output types helps you streamline your content operations. Creating deliverables and updating content is much faster when output types are standardized.

Using standardized output types, everyone—regardless of team, location, or job title—knows what content to include, in what order, and for what purpose. Reused content can be built right into the output type. This approach reduces the amount of time content creators must spend searching the repository for reusable content. It also reduces the risk of creating redundant content.

Automate Publishing

Using standardized output types, you can automate the publishing process. This process includes assembling content components, mapping them to the right stylesheets, and producing a deliverable (or many deliverables).

When you use standards to create content components, you do so without a particular format in mind. Anything you write, draw, or record can be published into any number of web page designs, presentation layouts, or chatbot windows.

In a standardized environment, design and formatting occur automatically during the publishing process. Simply define a standard output type, map it to a design, and click *Go*. (Okay—the process *might* be a little more involved than that. But you get the idea.)

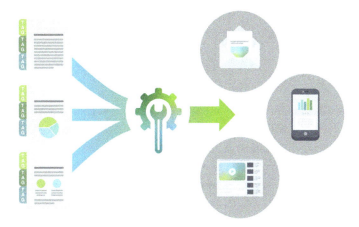

Prepare Content for AI

We believe that AI will one day help us publish highly personalized experiences without needing a standardized output type to guide the experience. In that future, AI systems will be able to deliver the right content to the right customer directly from our bucket of content components.

When AI becomes more commonplace, the concept of an output type for digital content might even begin to fade from our collective consciousness. For now, however, we still need standardized output types. In fact, they are one of the tools we can use to train today's AI systems.

From Static Delivery to Dynamic Delivery

In this book, we focus on delivering personalized experiences at scale. However, we know that no company can jump instantly from "no personalization" to "all personalization, all the time." You will likely continue to deliver at least some non-personalized content for some time.

During your transition toward personalization, standards become even *more* important. When all your content follows the same standards, it naturally flows together in a cohesive way. Even without more-specific personalization steps, such content delivers an experience with a more personalized feel.

Customers often ask us how to prepare to write in a componentized environment. Standardizing output types is one of the first things that we recommend. Regardless of your tools or authoring environment, you can organize all outputs of the same type in the same way.

That way, if a customer browses through your content and lands on a PDF manual, that manual will feel like a logical part of their journey.

A note about PDFs:

Even though we like to say that PDFs are where good content goes to die, the reality is that most company content still relies heavily on this type of deliverable. We hope that after reading this book, you'll be able to implement standards in all five content dimensions and lead the change management necessary to move your company towards personalization at scale. When that happens, your content might just break free from PDFs at last.

As you begin standardizing your output types, be sure to consider standards for three types of content delivery: static, curated, and dynamic.

- **Static outputs** deliver the same content to everyone.
- **Curated outputs** provide a certain amount of customization before being delivered to the customer.
- **Dynamic outputs** provide personalization at the point of delivery, by enabling the delivery system to match the content to the customer.

Static Outputs

Static outputs use the traditional print publishing model: Content is created, assembled, and published. The result is the same for all customers. It is the one-size-fits-all output type.

Static outputs have their place, even in a world of personalization at scale. For example, it will be a long time before the world's regulatory bodies allow pharmaceutical companies to deliver product labeling in a personalized, dynamic way. Until then, those companies must print all the required information in the required order to comply with various regulatory requirements. Every customer who gets a box of medicine also gets the printed label, which contains the same content in the same order as every other box. There's no difference in the content for a customer who has an extensive medical background and a customer with no medical experience.

The standardization best practices that we describe in this book apply equally to static, curated, and dynamic outputs. In fact, static outputs benefit greatly from standardized content:

- Standards enable content to be reused.
- Separating content from format enables multichannel and omnichannel publishing from a single source of truth.
- Content is created, reviewed, published, and retired in the most efficient way.

Most important, you can provide a better customer experience when your content is consistent across all output types, whether content is delivered as static, curated, or personalized output. Standardized content makes your static outputs easier to find, easier to use, and more seamless for customers who access more than one piece of content.

Curated Outputs

Curated outputs enable you to produce slight variations of the same source content. Curated outputs take static output and add two layers: tagging and filtering.

The difference between curated outputs and personalized experiences is that in the former, the content creator controls the curation. The content of each variation is the same for every customer who accesses that variation.

> Regina once worked with a technical communications team for a company that developed energy-data-management software for public and private utility companies. No matter how hard the company tried to make an off-the-shelf version of their enterprise application suite, every utility wanted their own custom experience. Some utilities wanted custom documentation for their custom implementation, and they were willing to pay for it.
>
> The technical communications team maintained multiple sources of truth that they delivered in long-form PDFs and online help systems. Every time the company updated the product, content creators had to go through both the common content *and* every customer-specific variant to make the same updates—twice, because they used one tool to create the PDFs and another tool to create online help. Reviewers had to review changes in every variant. Approvers had to sign off on every variant. Want to guess what translators had to do, if the customer purchased documentation in multiple languages?

> Part of Regina's role was to develop standards for how to provide curated outputs for each customer from a single source of truth. By the end of the project, the team had a single source of truth and a set of tags to use to quickly assemble customer-specific outputs.

To create curated output types, you need three things:

- Variables
- Conditions
- Publishing profiles

Variables

A variable is a placeholder that you insert in the body of the content, in place of specific words. When you publish the content, the variable is automatically replaced by the words that the variable represents. Use variables for content reuse when something small, such as a single product name, needs to be customized.

You can use variables as placeholders for all kinds of content. Companies often use variables for product names, model numbers, and customer names. If you've ever received an email (or letter) that started with "Dear FirstName," then you've seen what happens when variables go wrong.

Content creators also use variables as placeholders for information that they are uncertain of while creating content. Product names, for example, are often subject to change right up until the product is released. Rather than postponing their tasks, content creators can simply use a variable in place of the product name. When the name is finally official, they need only update the variable value, rather than performing a massive search and replace throughout all product-related content.

Conditions

Conditions provide a basic level of personalization by enabling authors to generate several variations of output from the same content. In some ways, the automated delivery of personalized experiences is simply conditions used on a massive scale.

In curated output, content creators apply conditions to content as they write. When the content is published, those conditions tell the publishing engine which content to include and which content to exclude from the output.

Conditional tagging is a fundamental way to personalize content. When content delivery is static, conditions permit the delivery of multiple variations without rewriting. When content delivery is dynamic, conditions can be processed on the fly, enabling mass customization of content. Conditions enable you to custom-tailor content to specific readers, based on what you know about them.

Publishing Profiles

Many organizations manage dozens of variables, each with many more than two values. Conditions also have a way of growing, with some teams managing 100 or more conditions. When you have many variables and conditions to manage, use a publishing profile.

Publishing profiles provide a way to group variables and conditions so that the content creators don't need to select each one individually every time they publish output.

Curated Output Example

A bicycle-tire company provides a manual that guides customers through various maintenance and repair tasks. Some customers have mountain bikes; others have beach cruisers. The instructions for fixing a flat tire are the same for both types of bike. However, the company provides a variation of the manual for each bicycle type to give customers more confidence in the instructions.

To simplify the process, the company's content creators use the variable placeholder `<bike_type>` instead of writing the word *bicycle*:

1. Turn the `<bike_type>` upside down.
2. Spin the tire so that the puncture is easy to reach.

To publish a mountain bike manual, the content creators select **mountain bike**. The publishing process automatically replaces every instance of `<bike_type>` with that type:

1. Turn the mountain bike upside down.
2. Spin the tire so that the puncture is easy to reach.

To publish the beach cruiser version of the manual, the content creators select **beach cruiser**. The publishing process again automatically replaces every instance of `<bike_type>` with the specified type:

1. Turn the beach cruiser upside down.
2. Spin the tire so that the puncture is easy to reach.

The company decides to make the content even more relevant to customers by providing different illustrations in each type of manual. Mountain bike owners will see mountain bike illustrations, and beach cruiser owners will see beach cruisers. To do so, the content creators add two illustrations to the content at each location that should include an illustration.

They tag the first illustration **mountain bike.** They tag the second illustration **beach cruiser**.

When publishing a manual, if the content creators select the **mountain bike** condition, the output includes all illustrations tagged **mountain bike**. Any illustration tagged **beach cruiser** is not included. To publish the beach cruiser version, the content creators simply select the **beach cruiser** condition instead.

The company can also create a mountain bike publishing profile and a beach cruiser publishing profile that both replaces the variable and specifies the condition for the appropriate bike type. Customers can visit the company's support site and select the version of the manual that matches their needs.

Curated Outputs Are *Not* Personalization

A curated output provides a rudimentary personalized experience. But it is *not* true personalization.

The substitution of a variable or application of a condition (like the one in our example) happens long before the customer sees the content. The content creator must still publish multiple versions of the output. The content delivery system does not match customer data with the potential variables or conditions to automatically provide the right content for the customer.

We call these outputs *curated* because the company makes the decisions about which content to include and in what order. The customer is not involved. They are getting a guided tour of a prebuilt house rather than choosing their preferred model.

Dynamic Outputs

Dynamic outputs are where personalization truly happens.

Unlike static or curated outputs—both of which rely on content creators to build the content they assume the customer wants to see—dynamic outputs assemble the content at the time the content is delivered. To do so, these outputs require much more sophisticated layers of tagging, automation, and filtering.

Dynamic outputs are relatively new to the publishing scene and require a lot of coordination, collaboration, and content engineering. This is one of the reasons why so many organizations struggle when beginning a personalization initiative.

Dynamic Output Requirements

For dynamic outputs to work:

- Content must be standardized.
- Content metadata must be consistent.
- Output type must be consistent.
- Customer data must be available.
- Rules must match content to customer.

Matchmaker, Matchmaker

Dynamic outputs can match content to customers only if you know about your content and your customers. And how much customer information you may use is limited by your technology, your policies, and various regional, national, and international laws and regulations.

But who knows your content better than you do?

The Customer as Curator

Dynamic outputs shift some of the curation process from the content creator to the customer in need of the content. Instead of the organization making all decisions about what to include in which output, the customer influences what content they see and in what order they see it.

Goodbye guided tour, hello personalized floor plan.

Customers may or may not be aware that they are helping create their own journey through your content. What they do notice is that the content they see is supremely relevant to them.

Regina consulted on a content strategy project with a support and operations team at an international payments-solutions company. This project involved all the usual efforts to analyze, componentize, and standardize content. The team worked hard to segment and tag the content to a fine level of granularity so that it could be matched to each customer's skill level, location, computer type, software version, permissions level, and attempted task.

What made this project unusual for its time was that the customers were all internal. The business recognized that if it could provide employees with a personalized experience, they could save tremendous time and money and mitigate risk of error.

We've seen a lot of progress in this direction from customer support organizations in a number of verticals. Support often has the advantage of knowing exactly who customers are, what products they own, and how much experience they have. When content is properly chunked and semantically segmented, support organizations can personalize content in a way that even marketing can be hard-pressed to beat. That's a massive gain, both for customers (who get the information they need faster than ever) and for companies (who can provide more-effective self-service options and earn points with customers). Boiling content down into the standardized, mix-and-match snippets that chatbots and voice assistants need to serve customers can be an investment, but it's one that's already proving worthwhile.

The Role of the Content Creator

The customer might determine their journey through your content, but you can't simply turn them loose in your content-component database to find their way alone. Rather, you need to set guidelines and create signposts so they know they're going in the right direction.

For dynamic outputs to work well, you must set up the rules that indicate which content to match to which customers, products, and situations. This type of mapping can become complex, as you incorporate many layers of information about your content and customers.

The personalization paradox is at work once again. To provide a personalized experience, you need to provide a high level of standardization to your delivery systems. These systems must be able to match customer data to content and serve the right content at the right time.

How to Standardize Output Types

To standardize output types, you need to define a content model for each type. A content model is a blueprint for content. It defines standards such as which components to include, where to include them, and whether they can be personalized.

To standardize output types:

1. Identify output types.
2. Develop content models.
3. Configure templates.
4. Connect formats.

[1] Identify Output Types

To determine the types of output you want to use, start by thinking through the purpose of each "package" of delivered content. You can use the sample output types from the beginning of this chapter to get started.

This step is absolutely crucial if you've never developed a unified content strategy or if you've worked primarily in unstructured content. Teams that have been through multiple authoring tools, content management systems (CMSs), reorganizations, and mergers or acquisitions also tend to have a lot of inconsistency in how content is assembled and delivered.

Standardizing your output types gives you an opportunity to clean up your content flow. And that enables you to reap the benefits of standardization even when you have no plans to personalize your content.

[2] Develop Content Models

A content model defines:

- A name for the output type, such as User Guide or Product Web Page
- The components to include
- The order of the components
- Mandatory content
- Optional content
- Reused content

- Where to add unique content
- Which content can be personalized (made different for different customers)
- Which content cannot be personalized (remains the same for all customers)

A content model does not define the design or the look and feel of published experiences. It is the blueprint of the house, not the house itself (with all its paint, flooring, and doorknobs in place). A housing developer can build many houses based on one blueprint. Likewise, if you adhere to standards in all five dimensions (output type, word, sentence, paragraph, and component), you can use the model to output any number of deliverables with an endless variation of content.

The separation of content from format is one of the most powerful aspects of structured content. Because the content is created according to standards and the format is applied during publication rather than during content creation, you can publish the same content to multiple output types and formats. For example, you can publish a user guide as both PDF and HTML. The same content is assembled according to the same content model but has a different design and format depending on the output type.

Oh, and those designs and file formats are also—you guessed it—standardized.

[3] Configure Templates

After you develop content models, you can configure templates in your CMS to enforce standards and support content creators:

- Provide an outline for each output type
- Prevent authors from deleting required components
- Enable authors to delete optional content
- Provide boilerplate text and headings
- Provide reused content
- Provide appropriate metadata

Templates do not define what the content looks like when it is published or released. The look and feel of the output are defined by stylesheets that transform your format-free, standardized content into its final format.

[4] Connect Formats

Output models define the organization of your content but do not dictate what the content looks like when delivered. That look and feel are defined by the *output format,* which defines design elements and functionality, including:

- Fonts
- Cover design for PDF and print outputs
- Color palette
- Navigation
- Branding

If the content model is the blueprint of your house, the output format lays out the finishing touches: paint colors, roof tiles, exterior siding, and so on. When you standardize the content model, you can use the output format to personalize each output.

Example: White Paper

Suppose your company publishes white papers. To standardize a White Paper output type, you begin by creating a content model that contains the following:

- One introduction
- Two case studies
- Three pairs of problems and solutions
- One executive bio (optional)
- One summary

The company asks you to make the white papers available as downloadable PDF files. And you sometimes need to publish papers directly to the company website.

You can develop several output formats for your PDF white papers:

1. Shades of brand-specific blue, portrait orientation, a light-blue gradient background behind the case studies, and the summary on the last page
2. Shades of brand-specific orange, landscape orientation, no special background for the case studies

3. Shades of brand-specific green, portrait orientation, the summary on the first page next to the introduction, both case studies on the last page, and no executive bio (even if one is available)

You can also develop different designs for your HTML white papers:

1. Lots of graphics, big hero images, and one infinite scrolling page
2. Few graphics, a left navigation menu, a click-to-view-next-page button, and no executive bio

You write and store each part of your content model as an independent content component. You can assemble and publish these components to any of your specified formats, at any time. Each component is format-free; the formatting is applied at delivery.

In this way, the same content can look dramatically different, depending on the chosen output. Yet completely different content still provides consistent branding. And that consistency creates a cohesive reading experience for your customers.

Summary

Standardizing output types is the first step in scaling content personalization. Standardized output types also improve the findability and usability of your content. Regardless of whether outputs are static, curated, or dynamic, standardization prepares you to deliver personalized experiences at scale.

CHAPTER 6
Managing Terminology: Standardizing Words

To mix and match content chunks on the fly at the point of delivery, you need standardized terminology. Otherwise, the same concept or thing ends up being named or described using different terms. And that prevents the easy flow of content, increasing the likelihood of customer confusion.

When you standardize terminology, every piece of content you create uses the same term to describe the same thing. You also gain control over which terms are allowed and which terms are prohibited.

We call this process *terminology management.* Terminology management helps to ensure that the right word or phrase is used at the right time, in the right context, with the right meaning.

> Have you ever decided to switch out your showerhead? If you live in the United States, the task is fairly simple because all showerheads in the country follow the same standard (ANSI standard ASME A112.18-1-2018, in case you're wondering).[1]
>
> Whether you want a massaging showerhead or a one that makes you feel like you're standing under a waterfall or in the rain, you can get exactly what you need in a matter of minutes, thanks to standardization. Without it, you wouldn't be able to mix and match choices so easily.
>
> Standardizing terminology does the same thing for your content.

---

[1] *ASME A112.18.1-2018: Plumbing Supply Fittings* (Fowler 2018)

Why Manage Terminology?

There are many reasons to manage and standardize terminology before you attempt to deliver personalized experiences. Here are a few:

- Successfully provide personalized experiences at scale
- Improve readability
- Lower the cost of developing and editing content
- Lower the cost of translating content
- Speed time to market
- Enforce legal compliance

Enable Personalized Experiences at Scale

To personalize content at scale, you need to create small, reusable content components that can be mixed and matched to provide unique outputs. When you write a component, you might not know all the places it will eventually be reused. And you might not know which other components will mix and match with the one you're writing, either.

In addition, components can be—and often are—written by more than one person. These content creators might not even be part of the same organization.

For example, sales and marketing might collaborate on content components that are used in a case study. If those content creators use different words to mean the same thing, the customer can easily be confused once those components are combined. Mismatched terminology can even create marketplace confusion, ultimately eroding trust in your brand. To create a successful experience, everyone needs to use the same terms across all components.

Improve Readability

Regardless of whether your company produces software, pharmaceuticals, or children's toys, consistent terminology makes your content easier to understand. This is true for all types of content: technical, marketing, sales, services, training, and so on. Using the same terms in the same ways sets a reading expectation. As a result, customers spend less time assimilating words and more time learning how to do whatever it is they need to do.

Imagine four content creators are collaborating on a book about dog care. Each completes content about one topic. The topics together form the final output.

- Jonathan is responsible for content about feeding.
- Pidge is responsible for content about grooming.
- Ben is responsible for content about exercise.
- Deb is responsible for content about training.

Without further ado, each writer gets to work. When they finish their topics, they meet to compare chapter titles:

- Feeding Your Canine
- Grooming Your Dog
- Walking Your Pooch
- Training Your Puppy

Because they did not manage their terminology, each writer used a different term for *dog*. All four terms are accurate—but they are completely inconsistent.

Trying to personalize content without standardizing terminology can result in confusion or, even worse, failure. Standardizing terminology makes your content more consistent, and consistent content is easier to read.

Lower the Cost of Content Creation

Standardizing terminology speeds up content development and editing, which in turn lowers the cost of creating your content.

Rather than searching for a different word to say the same thing, content creators just use the approved term. When the same words are used to mean the same thing each and every time, editing also becomes easier, quicker, and less expensive.

Reduce Translation Costs

Standardizing terminology lowers the cost of translation. Remember that "in the right language" is part of the definition of personalization. Translating content was one of the earliest forms of personalization, dating back thousands of years to a time when both source and translations were written by hand.

Content is translated in units called *segments*. A segment can be a sentence, a phrase, or even a single word. When you pay to translate content, translated segments are stored in a database called *translation memory (TM)*.

Because TM contains everything you've already translated, when you have new content to translate, you pay only to translate new segments. Any segments that exactly match those already in the TM do not incur a charge. Anything less than a 100-percent match is called a *fuzzy match*. These incur a reduced rate. The reduction percentage is a matter of negotiation, but usually depends on the percentage of fuzziness. The fuzzier the match, the more you pay, and payment for any unintended fuzzy match is money wasted.

So if your content creators say the same thing the same way every time, you can save a great deal of money on translation. But if content creators get creative and use different segments—even slightly different—to say the same thing, translation can get *very* expensive *very* quickly.

Table 6.1 – 37 ways to write one simple instruction

Click	Press	Select	Tap	Hit
Click Okay	Press Okay	Select Okay	Tap Okay	Hit Okay
Click on Okay	Press on Okay		Tap on Okay	
Click the button	Press the button	Select the button	Tap the button	Hit the button
Click the Okay button	Press the Okay button	Select the Okay button	Tap the Okay button	Hit the Okay button
Click on the Okay button	Press on the okay button		Tap on the Okay button	
Click OK	Press OK	Select OK	Tap OK	Hit OK
Click on OK	Press on OK		Tap on OK	
Click the OK button	Press the OK button	Select the OK button	Tap the OK button	Hit the OK button
Click on the OK button	Press on the OK button		Tap on the OK button	

Take a common task: activating an onscreen button. Table 6.1 illustrates 37 ways to write that one simple instruction.

If "Click Okay" is in your translator's TM, none of the other terms in this table are a 100-percent match. All of them will incur a fuzzy-match charge of some type. Furthermore, conscientious translators query the use of inconsistent terms. This query-research-answer process costs additional time and money.

You can see how quickly standardizing terminology can help you save money on translation!

Speed Time to Market

Standardizing terminology across all the organizations in your company enables you to take advantage of automated tools and processes. Automation speeds time to market.

Even without automation, using a shared, consistent set of terms across all organizations can reduce time to market. For example, having an agreed way to refer to things can shorten review cycles. Reviewers don't need to parse multiple terms that describe the same thing to decide if each is accurate or whether the different usage has a valid reason.

We also find that when content creators use consistent, agreed terms when they communicate with subject matter experts, the conversations are more productive. Less time is spent on definitions and making sure that everyone is talking about the same thing.

Enforce Compliance with Legal Trademarks and Service Marks

If one thing drives legal departments berserk, it's the incorrect use of trademarks, registered trademarks, service marks, and other legal demarcations. An incorrectly used trademark can violate the terms of that trademark. That's why legal departments get so picky about trademark usage. According to Trademarknow.com:[2]

"Since trademarks are issued to preserve distinctiveness, anything diluting the meaning of a mark can be grounds for cancellation. Not only do such instances result in lost rights, but they may also require businesses to either undergo expensive rebranding campaigns or forgo a particular product or service entirely."

Managing legal terms helps to enforce correct usage, which protects your trademarks.

---

[2] "6 Frustrating Ways to Lose Your Trademark Rights" (Potts 2017)

What Terms Should You Manage?

A managed term is a word or a set of words that have a meaning unique to your company or brand. Don't expect to manage every term in your content.

Sometimes deciding which terms to include and which to leave out can be difficult. We recommend that you focus on the following categories.

Terms with Legal Implications

When used incorrectly, terms with legal implications can open your company to a potential lawsuit. Correct use of such terms is crucial. Your legal team probably keeps close track of these terms, particularly if your company operates in a highly regulated industry.

Product and Brand Names

Using product and brand names correctly is near and dear to the hearts of all product managers and marketing directors. Strict adherence to these names helps establish your product and brand in the marketplace. If you are inconsistent with product or brand names, customers can become confused and content seems sloppy. Plus, you can accidentally create trademark or copyright implications. If disagreements occur over which product or brand names are correct, check with your legal team.

Neologisms and Portmanteaus

Have you noticed that companies love to make up words?

We've created entire categories of products using made up words: eBook, ePub, eNewsletter. Just add a lowercase "e" or "i" to any word … someone is sure to be using it. Or to quote one of our favorite Calvin and Hobbes cartoons, "we verb nouns." We no longer look something up on Google, we Google it. We don't ship a package using FedEx, we FedEx it. We don't send out messages, we message one other.

We also create entire languages (or at least dialects). In addition to Tolkien's Elvish and Rodden-bery's Klingon, we have LOLCat and Doggo. These languages have quickly infused the marketing content of companies whose brand voices leverage the latest memes and themes.

When you use words that aren't in any dictionary, making sure everyone uses them in the same way becomes even more important. If your content needs to change quickly to keep up with the latest internet language, then you simply must manage those terms.

Prohibited Terms

Every company has a list of words it wants content creators to avoid. You might decide to prohibit a term for a variety of reasons:

- Clarity
- Consistency
- Comprehension
- Brand
- Translation

Keeping a list of prohibited terms is important. Equally important is selecting at least one preferred term for each term that you disallow. Telling people which terms they cannot use, without giving them an approved replacement, is annoying at best. At worst, you risk another prohibited term being selected in its place.

For example, our company name is officially Content Rules, Inc. But when we write about the company, we drop the "Inc." and use Content Rules. To make sure all our content creators do so, we prohibit "Content Rules, Inc." and specify "Content Rules" as our preferred term. We also prohibit variants of Content Rules, such as ContentRules (which we see often, though never in our own content) and Content-Rules.

Any time companies merge, content creators are usually asked to remove all references to one or both original company names, depending on what the new, merged company is called. In this case, prohibit the old company name or names and make the new name the preferred term.

Terms That Promote Inclusivity—or Add Bias

Everyone has unconscious biases that emerge in our writing. These biases come from our personal experiences with our families, countries, regions, religions, relationships, education, and so on.

The internet, too, is inherently biased. Stereotypes tend to be perpetuated by ideas that are not challenged or questioned. For example, open a browser, type "Why do women always," and see how the browser auto-completes that sentence. What you find might (or might not) surprise you.

Managing terminology helps to prevent these biases from making their way into your company's content. It also helps each of us become more aware of how we unconsciously (or consciously) perpetuate stereotypes or assumptions. In this way, managing terminology improves both our content and ourselves.

Here are some examples of biased words that most companies now exclude from their terminology:

- Master/Slave
- Whitelist/Blacklist
- Man hour
- Chairman

Terms with Unusual Capitalization

Companies have a tendency to get creative with capitalization. They might make proper nouns of everyday words in an attempt to emphasize their importance. Or they might capitalize random letters as a part of a branding attempt.

Whatever the reason, if you capitalize too many words, customers start ignoring them. Capitalization-happy text can actually create confusion about which terms are brand- or product-related terms. Plus, capital letters are distracting. Use them only when absolutely necessary.

Another reason to avoid unique capitalization: Every time you use a common word in uncommon ways, you must manage that word to guarantee consistency. And that adds time and effort to your content creation workflow.

Terms with Multiple Variations

Many words that have multiple variations that are all accurate, according to your spellchecker. For example:

- email and e-mail
- bicolor and bi-color
- standalone, stand alone, and stand-alone
- fiber optic and fiber-optic
- check box and checkbox
- OK and okay
- login and logon

It's not just your spellchecker. The Oxford Dictionary contains both *login* and *logon*. But Merriam-Webster defines only *login*. If you look up *logon*, the dictionary refers you to the page for *login*. See what happens when we make up words?

To make it even more confusing, different dictionaries provide different spellings for the same word. For example, Merriam-Webster allows Web site[3] as two words with a capital W. The Oxford Dictionary uses a single word, all lowercase, website.[4] In this situation, looking up correct usage yields different results.

Unless you have standardized on a corporate dictionary, writers can be confused. Pick one word (actually, it's best to pick one dictionary) and use it consistently. Make sure the term is managed so that everyone uses the same version.

Another type of variation involves punctuation. Terms such as "e.g." and "i.e." often create some confusion when it comes to how many periods to include and where to include them.

What Not to Manage

When it comes to managing terms, steer clear of one **big** category:

Common words that are used in a common way.

There is no need to manage a term that is simply a word (or a word cluster) with nothing special about it. Doing so is the equivalent of using six fonts just because you can.

How to Manage Terminology (and Standardize Words)

Terminology management requires you to develop standards for both storage and retrieval.

[1] Locate and Gather Your Terminology

To standardize your terminology, the first thing you need to do is locate it. Many companies that we work with start by telling us they don't have terminology. But *every* company has terminology.

---

[3] http://www.merriam-webster.com/dictionary/website
[4] http://oxforddictionaries.com/us/definition/american_english/website?q=website

- Start with your product names, trademarked words, and service marks. If your company sells a product or service, we can guarantee that you have at least one product or service name.
- Next, look in any corporate style guides or glossaries your company has. Tip: Check with your localization team to see what terms they include in the glossaries they send to translation.
- After you gather up your terms, determine which words need usage guidelines, which ones to allow or prohibit, and so on.

You can also use software to help identify meaningful terminology. Some software offers a *term harvest* capability. During this process, you provide content for the software to ingest. It runs analytical algorithms to identify terms that appear to be used in certain ways or to fall into the categories we mentioned earlier. The software then exports a list of these terms, along with a score indicating the likelihood that they should be managed. The list still requires human expertise to analyze and finalize the right terms. However, such software can give you a significant head start if you have a lot of content, creators, or silos.

[2] Determine Terminology Storage

You can store terminology in two ways: flat file storage or a terminology management system.

Flat File Storage

A flat file is a storage system that has no inherent hierarchy. It is simply a file that you can read, add to, and search within. Depending on the tools you use, a flat file can contain internal or external cross-references. But it is *not* a database.

More than 80 percent of the companies we work with use a flat file system to manage their terminology (that is, if they manage their terminology at all). The two most common file types are Microsoft Word and Microsoft Excel. Terminology is typically stored in one or more tables; rarely have we seen companies use numbered or bulleted lists, which are too unwieldy even for the smallest terminology bank.

A flat file can be convenient. It is:

- Easy to create using only standard tools
- Inexpensive, requiring no investment in new software
- Simple to use and modify

However, it might also be the worst method for storing terminology. It is:

- Difficult to find terms unless you know exactly what you are looking for
- Increasingly painful to manage and maintain as the number of terms increases
- Inefficient
- Potentially difficult to control

Terminology Database

A database—a collection of organized, categorized data—is a much more powerful way to store terminology.

A database that is designed specifically for terminology is called (surprise!) a terminology database, or *termbase*. A termbase can be in one language or multilingual. Multilingual termbases are common in localization and translation, and you can choose from many multilingual termbase tools on the market.

Using a termbase, you can:

- Categorize terms easily and extensively
- Search for one or more terms based on a variety of criteria
- Act on search results in a variety of ways
- Use different applications to work with terms
- Gain better control over who can access terms, what each person can do, and so on
- Store and manage multilingual terms in one place, keeping them in sync

However, termbases do have potential disadvantages:

- The software costs money.
- Databases are more complicated than flat files.
- Someone must administer the database.
- There is a learning curve to using a termbase.

If you have more than 100 terms, using a termbase is the best choice. If you have only a few terms, a database is probably more than you need. But if your content involves translation, always use a termbase, regardless of how many terms you need to manage.

[3] Determine Terminology Retrieval

You can use pull or push methods to retrieve terms from storage.

Pulling Terms

A pull method for terminology retrieval is just as it sounds. When you need information about a term, you must pull that information from your database or flat file. With pull methods, content creators need to:

- Think about the term they want to use
- Decide whether the term is managed
- Locate the terminology (by opening the flat file or launching the database interface)
- Search for the term
- Look up information about the term, such as whether it is permitted or prohibited and whether special usage rules exist
- Use this information to complete their writing task

If you have more than 50 terms, this process can be inconvenient at best and a tedious waste of time at worst. In most environments, content creators are under tremendous pressure to deliver content as quickly as possible. We continue to be asked to do more with less: less time, fewer people, and so on. How often do we have time to stop what we are doing, think about a particular word, look it up, and act on it?

The answer is rarely, if ever. While we all strive to deliver the best possible quality in our content, rarely do we have the luxury to look up words in a terminology list. Using a database that centralizes the information we need about our terminology is certainly faster than hunting down every piece across multiple resources. Still, pull methods require time that we rarely have.

Pushing Terms

With push methods, terminology information is automatically pushed to content creators during the writing or editing process. You don't need to stop what you're doing, go through the whole look-up process, and act on what you find. Instead, the information is delivered to you with the touch of a button. (Note: Push methods work only with a terminology database.)

The best push technologies deliver the information you need right inside your authoring tool, whenever you need it. For example, when you use a prohibited term, a push tool can highlight the term, tell you it is prohibited, and suggest the preferred term (or terms). You don't need to find your term list or even leave your current application. You don't need to look anything up. All the information you need is delivered to you instantaneously.

More advanced push technologies are based on complicated, highly configurable natural language processing (NLP). Some tools even understand parts of speech.

> Suppose your company decides to standardize the term *verify*. You make *verify* the preferred term and prohibit (or deprecate, a term often used within terminology management systems) all synonyms, such as *check*, *ensure*, *make sure*, and so on.
>
> A content creator writes the sentence:
>
> *Check that the lights flash once per second.*
>
> If you have push terminology management technology in place, the software automatically flags the term *check*, notifies the creator that the term is prohibited, and suggests the term *verify* instead. Simple enough.
>
> Later in the document, the content creator writes the sentence:
>
> *Pay for the software update by sending a check to the following address:*
>
> Here, the term *check* is used as a noun rather than a verb. Sophisticated NLP engines recognize such differences. Therefore, the software doesn't flag the term *check* in this sentence.

Push technology is neither cheap nor simple. However, it is the gold standard for managing terminology. The gains in efficiency, time, and dollars saved in content development and translation add up quickly. Most companies achieve ROI for a push terminology management system in about a year. After that, the savings multiply—and content quality improves exponentially.

[4] Implement Terminology Management

Whichever method you choose—flat file or database storage, push or pull technology—you need to set up your terminology management system. You need to ensure that everyone involved with content creation has access to it. And you need to train those people on how to use it to maintain consistent terminology throughout all your content. Perhaps most important, you need to ensure that everyone understands why that goal is important and make it a priority in the writing and editing processes.

Summary

Words make content fit together—or fall apart. Standardizing terminology helps to ensure that you can use, reuse, and deliver any content component to any customer at any time and that all components fit together smoothly.

CHAPTER 7

Managing Grammar and Style: Standardizing Sentences

For one reusable component to flow into the next, your sentences must follow the same rules of grammar and style. If you do not standardize your sentence structures, you risk mismatch and confusion when you try to deliver personalized experiences.

Plumbing and Electrical

In home building, the standards that govern plumbing and electrical systems define how various fixtures can be combined. For example, the plumbing system in a home is made up of a main that brings water from the source, pipes that carry water to fixtures, pipes that send wastewater to a sewer or septic system, and so on.

Certain rules must be followed when you put together a plumbing system. Some rules are based on municipality codes, others on the laws of physics. Similarly, grammar and style rules govern how words are combined. For the system (your content) to work, every sentence needs to follow the rules.

Sentences connect your content components to one another. Component-based content typically doesn't use transition sentences to "glue" those components together; instead, component titles (headings) do that work. But if your sentences do not fit together like one pipe into the next, your content isn't going to flow well. And without that flow, the customer experience is going to stink like a broken sewer pipe. Trust us on this.

Grammar and Style are Fluid

Language is not static; it constantly changes. Each year the English language adds and removes words or changes their definitions and usage. Jargon, idioms, and colloquialisms are in constant flux. Generational slang gives new meanings to sayings we've used since we were teenagers. Even a company name can become a verb, if the company becomes embedded enough into current culture.

When it comes to sentence structure, though, the basics of English grammar change much less frequently. Style, on the other hand, is as constantly evolving as our words themselves. And you'd be surprised at how much of what we think of as grammar is actually about style.

To standardize sentences in our content, we need to manage both grammar and style.

Grammar

Most of us learned about grammar back in elementary school (sometimes referred to as grammar school). In technical terms, grammar is the study of how words are used properly in a sentence. Typical topics in the study of grammar include:

- Parts of speech (also called word classes), such as:
  - ► Nouns and pronouns
  - ► Verbs
  - ► Adjectives
  - ► Adverbs
  - ► Prepositions
  - ► Conjunctions
- Negation, which includes:
  - ► Using the word *not*
  - ► Contractions such as *isn't*, *can't*, and *couldn't*
  - ► Other negating words such as *never*, *nothing*, and *nobody*
- Sentence structure, which includes:
  - ► Word order
  - ► Dependent clauses
  - ► Imperatives

Why Grammar Matters

 Bad writing makes bright people look dumb.

—William Zinsser, author of *On Writing Well*[1]

Using correct grammar is important for many reasons:

- Accuracy
- Readability
- Ease of translation
- Impression of your brand
- Search engine optimization (SEO) for certain browsers

To mix and match content components to create a personalized experience, each component needs to be grammatically correct. Every component must also follow the same rules of grammar. If, for example, you mix past, present, and future tenses (without a plan or reason for doing so), your reader can become confused about what is happening now, what happened in the past, and what will happen in the future. To reuse and remix small chunks of content in a readable way, you need to standardize the grammatical rules that govern sentences.

Different Forms of English

There are many forms of the English language, including (but not limited to):

- U.S.
- British
- Scottish
- Australian
- Canadian
- South African
- Standard
- Global
- Colloquial
- Plain
- Doggo

---

[1] *On Writing Well* (Zinsser 1990)

Different forms of English use different spellings, pronunciation, and grammar. All your content needs to follow the same form of English. For example, you don't want to combine British and American spellings (such as *practise* and *practice*, *colour* and *color*). You need to pick one form and stick with it in all your company's content.

Our advice for selecting your form of English is to keep it as simple as possible. We suggest using Plain, Standard, or Global English for all your content. That way, everyone can read it. If you want to provide other forms of English in your content, you can localize components from your simple English source.

Grammar Rules to Consider

When you use standard, non-colloquial English, you have many grammar rules to consider: 10, 20, or even more, depending on whom you ask. Some grammar rules are simple. For example, always start each sentence with a capital letter and end with a full stop (period), question mark, or exclamation point. Always make sure that a singular subject and verb agree in number.

Many grammar rules involve word choice. For example:

- *Affect* versus *effect*
- *Their* versus *there* versus *they're*
- *Fewer* versus *less*
- *Many* versus *few*
- *Its* versus *it's*
- *A* versus *an*

Other grammar rules are more esoteric and cross into the realm of style:

- Avoid passive voice.
- Use the present progressive tense for current action.
- Use the present perfect for the unfinished past.
- And the most widely contested rule of them all: Use—or shun—the Oxford (serial) comma.

Agreeing on your grammar rules and ensuring that everyone follows them is vital to creating personalized experiences. Otherwise, you can easily end up with sentences, paragraphs, and components that do not work together.

For example, if you standardize on the present tense for current action, but some of your content creators use the future or past tense, personalized output will likely be confusing. What happens now? What happened earlier? What is yet to happen? Who can tell?

Style

As we mentioned previously, people often confuse style with grammar. Grammar defines technically correct ways to use words according to a specific language. Style defines choices that affect the way sentences are understood; it forms a bridge between words and grammar and between brand and voice.

Style comprises the intentional, specific, detailed decisions you make about how content creators write on behalf of your company. Following a consistent set of style rules helps make your content components interchangeable and their assembly seamless. It also helps content that is created across different regions and countries sound the same, so that it can be reused and delivered to customers anywhere.

Style assumes that you are choosing from grammatically correct options. After all, if your grammar is inaccurate, your style choices don't matter. Standardized style can't overcome grammatically incorrect sentences.

There are *hundreds* of style rules, including decisions such as:

- Avoid contractions.
- Use only contractions.
- Avoid future tense.
- Avoid Latin expressions.
- Avoid "s" in parentheses.
- Avoid possessives.
- Avoid pronouns.

Style is also where you support and enforce best practices such as global writing (e.g., always use a noun with *that/these/this*), legal considerations (e.g., always include a hazard statement if a risk of bodily harm exists), or regulatory compliance (e.g., every claim about a pharmaceutical product must be accompanied by this regulator-required language).

Because you have so many style choices to make, we suggest that you start by considering the big picture. What are your content goals? Which style rules or practices can help achieve those goals?

Thinking about your content from a goals-based perspective can help you decide which rules are useful and important and which just amount to busywork.

Idioms, Jargon, and Colloquialisms

Many companies adopt a colloquial form of English. Consider Figure 7.1, which is from a web page belonging to a real, but unnamed, camera manufacturer. The text on this page is *idiomatic*.

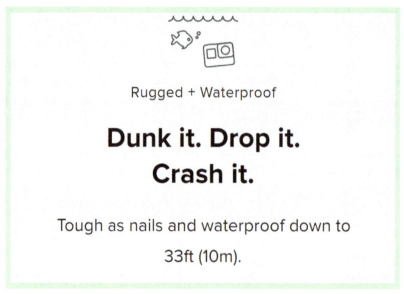

Figure 7.1 – Idiomatic text[2]

---

[2] Source: https://gopro.com/en/us/shop/cameras/hero7-black/CHDHX-701-master.html

Or look at this description of an online game (Figure 7.2). Words such as "hyper-real," phrases such as "palms of your hands," and instructions such as "Team up with friends to form a crew," are *jargon*.

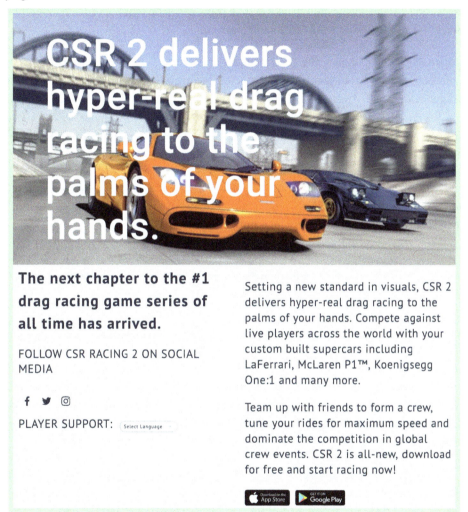

Figure 7.2 – Jargon[3]

---

[3] Source: https://www.zynga.com/games/csr-racing-2/

In many ways, using jargon is a good way to personalize content. Jargon usually appeals to particular demographics. In 2018, for example, the Oxford English Dictionary (OED) started an appeal called "Youth Words" in which the OED is attempting to catalog the fast-changing world of teen jargon.[4] What can be more personal than using a lexicon that a specific demographic embraces when providing content for that demographic?

Unfortunately, jargon, idioms, and colloquialisms have their drawbacks. Some of your content consumers might not have learned English as a first language, making colloquialisms extremely difficult to understand. And colloquial English is often impossible to translate. Because colloquialisms vary from place to place, age group to age group, and gender to gender, what works for one demographic can completely backfire with another. Therefore, using colloquial English as a grammar and style standard is risky business.

Content creators also comprise different backgrounds, ages, and cultures, each with different idioms, colloquialisms, and jargon. Even if your brand needs to use these forms of English, you might still need to write most of your content in Plain English.

When you need all your content to scale, avoid idioms, jargon, and colloquialisms. Instead, opt for simple English that is easy to read, understand, and translate. Weigh the effects of customized colloquial content against the scalability of simple, plain English.

> Not every piece of content that you create needs to follow the exact same set of standards.

> In general, marketing content is *emotional content.* It is meant to evoke some type of emotion. It might make us laugh; it might make us wonder. Most marketing departments want us to feel that they *get* us—that they understand and are speaking directly to us. Your advertising organization might also prefer to create unique, custom content that has a short (but powerful) life span.

> To accomplish this goal in specific situations, you can create multiple versions of content that contains colloquialisms: one version per demographic, using specific words that appeal to that demographic. This approach is fine as long as it is planned and intentional.

---

[4] "Appeals: Youth Words" (Allen 2018)

Style Guides

Numerous standard style guides are available for you to use. How you use them is up to you. You can use one style guide exclusively, use more than one, mix and match style guides, or come up with your own corporate style guide. Here are a few popular options.

The Associated Press Stylebook

The Associated Press Stylebook (often referred to as the AP Style Guide) is an English style guide that was originally created by American journalists working with the Associated Press. Even though the guide was originally written for journalists, over the years it has become one of the leading style guides for corporate communications and digital content. As of February 2021, the AP Style Guide is in its 55th edition.[5]

Topics include:

- Business guidelines
- Sports guidelines and style
- Guide to punctuation
- Briefing on media law
- Photo captions
- Editing marks
- Bibliography

The AP Style Guide is popular among marketing copywriters. Its origination as a newspaper style guide means that it guides writers toward concise, compelling sentences. Though formative concerns such as setting lead type and fitting content in the narrow columns of a printed newspaper are less important today than when the guide was first published, many companies still use the AP as their style guide of choice.

Microsoft Writing Style Guide (Formerly Microsoft Manual of Style, or MMoS)

The Microsoft Writing Style Guide is published by Microsoft. Although originally intended for the company's (and its vendors') content creators, it has become a standard style guide for many technology companies.[6]

---

[5] *AP Stylebook Online* (Associated Press)

[6] *Microsoft Writing Style Guide* (Microsoft 2018)

Much of the Microsoft Writing Style Guide is specific to Microsoft, such as the topic "Top 10 tips for Microsoft style and voice." However, many important topics are company-agnostic, including:

- Bias-free communication
- Chatbots and virtual agents
- Accessibility guidelines and requirements
- Global communications
- Responsive content
- URLs and web addresses

Tech companies often use the Microsoft Writing Style Guide to help ensure that their publications contain industry-standard styles and terminology. Software documentation particularly relies on it to help ensure consistent treatment when writing about user interfaces and common customer/ computer interactions.

Global English Style Guide

The Global English Style Guide (also referred to as Global English) describes best practices for writing English content that is most easily understood by non-native speakers.[7] This guide also contains rules for global-ready writing. The guide focuses more on style than grammar and includes examples to help authors broaden their understanding of a global audience. Sample topics include:

- Eliminating clichés, colloquialisms, and other ambiguities
- Punctuation and capitalization guidelines
- Simplification
- Explicit sentence structures
- Language technologies

Global English typically augments, rather than replaces, other style guides. It is particularly useful for global companies that translate content into five or more languages.

---

[7] *The Global English Style Guide* (Kohl 2007)

The Chicago Manual of Style

The Chicago Manual of Style (CMOS) is an American English style guide that has been around since the early 1900s. As of February 2021, it is in its 17th edition. CMOS is one of the most widely used writing style guides available.[8]

Val has had her copy of CMOS for more years than she is willing to admit. It is her go-to book for style decisions and includes topics such as:

- The publishing process
- Style and usage
- Source citations and indexes

CMOS is most heavily used in traditional publishing companies that produce hard copy and electronic books.

Custom Style Guides

Most large companies use one of the standard writing style guides as a starting point for creating their own custom guide. In addition to topics that focus on sentences and parts of sentences, custom style guides often include:

- Allowed and disallowed terms
- Acronyms and company-specific words
- Legal boilerplate wording, including caution and warning notices
- Preferred systems of measurement to use (imperial/metric)
- Index usage and style
- Glossary usage and style

A writing style guide is sometimes incorporated into a larger, brand style guide. A brand style guide might also contain information about:

- Logo usage
- Corporate colors and fonts
- Visuals
- Brand mission
- Formatting
- Tone and voice

---

[8] *The Chicago Manual of Style* (University of Chicago 2017)

The Problem with Too Many Style Guides

Providing too many style guides is just as bad as not providing any. Multiple style guides can confuse content creators (and customers) if their rules conflict with one another. We suggest you pick one style guide (or make your own) and use it across all your content.

Training People to Use a Style Guide

Many professional content creators were never taught how to use a style guide. They might have come to content from another field or were never mentored by an old-school editor.

The general rule is that if a content creator has even the *slightest* hesitation or question about what they are writing, they must stop and look it up. Over time, they'll memorize the styles and need to look up only things that have changed. (Such changes, of course, will be clearly and effectively communicated.)

If the content creator starts working on a new team, on a new project, or with a new product line—any new thing that comes along—they should skim through the style guide to see which rules might apply that didn't before.

We recommend that when you onboard new hires, contractors, or vendors, you don't just hand over your style guide (or list of style guides). Instead, take content creators through the guide and highlight the rules that are particularly important to your content. You might even consider creating a Top 10 list of the most important style rules that you want new creators to follow. That way, they know what to focus on first.

Remind them that their content will have a bigger impact than just one web page or document. To support positive, personalized experiences for customers, they need to write content components that flow well with others, no matter who wrote them or who manages them. Emphasize consistency. Rewarding them with chocolate can be an effective incentive, too.

The Dirty Little Secret About Style Guides

Unfortunately, most corporate style guides are created and managed manually, using a plain document or a spreadsheet: the most outdated and inefficient tools on the market. You might as well use a legal pad, pen, and three-ring binder.

As we saw with terminology, manually using a style guide is a pull process. You can maintain your style guide in a static file, which content creators must search through when they have a question. But here's the dirty little secret about storing style guides in documents and spreadsheets: No one uses them. That's right: No one. Maybe a new content creator will refer to a list for the first deliverable. But after that, no one has time to look up things manually. Most writers and editors don't have the luxury to use an inefficient, time-consuming, manual process.

Instead, make using your style guide easier for content creators by automating it, turning adherence from a pull process into a push process.

Automating Your Style Guide

We recommend using software to ensure adherence to your corporate terminology, grammar, and style rules. As with push terminology tools, push grammar- and style-management software can parse each sentence, flag grammar or style errors, and even correct the issue in many cases. And unlike humans, these tools don't get tired, feel rushed, or lose their place.

Another benefit of automated style guide applications is that they provide an objective way to correct content. As we know, content creators can feel judged by edits and corrections during content review. Software removes the personal tensions that can arise when a colleague proofreads content. (Although because software cannot catch *every* possible error in the English language, we recommend having a person spot check content after the software process is complete.)

There are three general classes of style guide automation:

- Built-in spelling and grammar checkers
- Editing software add-ons
- Enterprise content optimization software

Built-in Spelling and Grammar Checkers

Most of us are familiar with the red wavy lines that show up under our spelling errors when we work in office productivity apps like Microsoft Word or Google Docs. These applications also include grammar-check capabilities. (In Word, if you turn on real-time grammar corrections, you see wavy purple lines under the grammatical errors).

A built-in spell checker and grammar checker is the most basic type of style guide automation. And although you can customize a built-in spell checker to some degree, built-in grammar checkers typically are not customizable. Keep this in mind as you move to standardize your content so that it is personalization-ready.

Editing Software Add-Ons

The next level up from a built-in spelling and grammar checker is an editing software add-on. The most popular platform in this category is Grammarly, followed by ProWritingAid, Linguix, Textio, Sapling, and a host of others.

Most tools in this category offer free versions with limited capabilities. For a reasonable sum, you can usually upgrade to a paid version that extends the features. Editing software add-ons have features such as:

- Spelling check
- Grammar check
- Missing article detection
- Repeated word detection
- Weak adjective detection
- Plagiarism detection

Editing software add-on tools are definitely an improvement over built-in spell checkers and grammar checkers. They are good options for individual creators and individual pieces of content. However, they are not good options for a comprehensive collection of company content.

Enterprise Content Optimization Software

Enterprise-class platforms are the most feature-rich (and expensive) tools for managing spelling, grammar, and style. These high-end solutions also tend to include many features that are unavailable in standard software add-ons, including:

- Terminology management
- Extensive reporting
- Sentence reuse suggestions
- Multiple rulesets for grammar and style
- Multiple domains for terminology

These tools are highly configurable and run in dozens of content creation environments. You can use them to program your company style guide, making sure the rules in the tool match your standards. You can support multiple rulesets and multiple terminology domains for checking different types of content. For example, marketing content can have a more informal style than product support content, and legal content can have a much more formal style.

Enterprise content optimization software includes products from Acrolinx, Congree Language Technologies, Etteplan, and Writer.

How to Standardize Sentences

Creating grammar and style standards for enterprise content takes a combination of linguistic savvy, aesthetic awareness, and diplomacy. Just ask three people how they feel about the Oxford comma (or as it's known in the United States, the serial comma). You might never reach consensus—but if you begin to think the debate is meaningless, just look up lawsuits that have been won or lost based on comma placement.

We say this not to scare you, but to prepare you.

To standardize sentences:

1. Determine your base language.
2. Select a base style guide.
3. Determine your company styles.
4. Configure your software or create your style guide.

[1] Determine Your Base Language

Your base language defines your standards for grammar.

We recommend keeping the base language as simple as possible. Contrary to popular belief, American English is not an international standard. As we mentioned previously, we recommend using Plain English, Standard English, or Global English. You can easily localize from Plain English into American, British, or other forms of English.

[2] Select a Base Style Guide

Your base style guide gives you a head start when it comes to enforcing style standards.

Although using only one style guide is best, no single style guide covers every possible variation in style. Therefore, many companies use more than one guide. If you do, make sure you document which parts of which guide content creators should refer to as a single source of truth.

Back in the early 2000s, many companies used AP style for marketing content and CMOS or MMoS (or both) for technical content. As you begin delivering personalized experiences at scale, you'll find that you're able to provide a better customer experience if all your content follows the same set of standards. If you do you need to use different styles for different types of content, be sure to document those variations and apply them consistently.

[3] Determine Your Company Styles

Your company styles are the standards you develop when:

- Your style standard conflicts with your base style guide
- Your base style guide does not address your style standard
- You allow intentional variations or exceptions to style standards
- You promote deviations from the style standards for branding purposes

Document any company style standards for clarity and consistency.

[4] Configure Your Software or Create Your Style Guide

Incorporate all your style decisions into your content optimization software or style guide documentation.

Designate one or more people to manage the style standards. If you work with outside agencies or outsourced resources, ensure that those content creators and editorial teams have access to your style standards as well.

Summary

To create a scalable, personalized customer experience at scale, you must be able to seamlessly meld different content components. To mix and match components, your words must align and your sentences must adhere to the same grammar and style rules.

CHAPTER 8

Managing Voice: Standardizing Paragraphs

Paragraphs give your content a *voice*. That voice helps you connect with your customers.

Marketing organizations are familiar with voice. They typically develop your company's voice and ensure that it faithfully represents your brand across all customer-facing marketing materials.

Other content teams typically don't participate in creating the voice and may not even be aware that the company *has* a voice. But if the support, training, and documentation teams do not follow your brand's voice standards, your customers are in for a jarring experience when they access your content.

Many companies hire outside agencies that specialize in developing voice and tone strategies. These projects can take weeks, months, or longer. The result might appear to be as simple as a list of adjectives (or *voice attributes*) such as "adventurous, optimistic, warm" or "professional, experienced, wise." However, the process of defining a brand voice can be challenging.

 In its purest form, voice is how the relationship comes to life.
—John Caldwell, author of *Voice and Tone Strategy* (Caldwell 2020)

What Is Voice?

Voice is your company's personality. It is how your company represents itself in the world.

Is your company formal or casual? Is it established, old, and wise? Is it fresh, young, and disruptive? Does it move fast, anticipate trends, and change often? Or is it more deliberate, slower to change and acknowledged for reliability and security?

Large companies simply cannot provide individual human contact for every customer. But when you get your brand voice right—and when every piece of content follows those voice standards—you can help customers feel like they're getting that individual experience. That experience is the basis of the relationship between your company and your customer, for better or worse.

To provide personalized experiences at scale, you must create content components that can be assembled on the fly to reflect your company's approved voice.

Just as housing developers enable home buyers to control the aesthetic details of the homes they commission, we cannot impose one-size-fits-all documents on our customers when we deliver personalized experiences. We must give our customers an active role in choosing which content they see. If we do our jobs well, our customers don't even realize that they're making choices; they just know that they see what they need and like what they see.

Aesthetics

Your content's voice is akin to the aesthetic touches you choose to put into your home. Such aesthetics involve finishes, landscaping, and décor that give your house a certain feel. Here are a handful of popular aesthetics:

- Cape Cod
- Colonial
- Craftsman
- Mediterranean
- Mid-century Modern
- Ranch
- Tudor
- Victorian

Even houses that have an identical floor plan can have different aesthetics as the result of small details, such as doorknobs.

Craftsman   Victorian   Mid-Century Modern

Each doorknob does the same thing, but each presents a different aesthetic. The home buyer has the option to choose which knob to install throughout their new home. Likewise, they can select light fixtures, moldings, cabinet handles, window frames, and other finishing touches. Ideally, the home builder will offer such options via a select number of packages, each of which reflects a specific aesthetic.

As John Caldwell (author of *Voice and Tone Strategy* (Caldwell 2020)) says, "Think about how you are going to validate aesthetic choices to connect with customer desires. You can personalize a house with awesome styles that follow the standards. But if they don't resonate with what the customer wants or expects, they'll still leave the house. Emotions are in play, and emotions aren't rational. This is key to any strategy that employs voice."

How Paragraphs Express Voice

A paragraph is a collection of sentences that build on each other to convey information and to express voice. Customers might not be conscious of it, but they have certain expectations about paragraphs:

- Each sentence in a paragraph should build on the previous sentence.
- A paragraph has a beginning, a middle, and an end.
- Variety in sentence composition is needed to prevent monotony.

The words you use, the length and styles of your sentences, and how you combine them all come together to put your voice into the content. And your voice is what turns reading into an experience, rather than just a list of sentences.

Paragraphs also provide a visual indicator of where a piece of information begins and ends. The space between paragraphs helps customers skim through a block of content to find the information they want. When you standardize your components well, your customers know exactly which paragraph to go to if they are looking for something specific.

Have you ever read a novel in which the publisher left out the line breaks that indicate shifts in time, location, or scene? If so, you know how powerful the spaces around a paragraph can be.

> When one of Regina's favorite novels was converted to an eBook format, those line breaks were dropped. The story depends on many scene and point-of-view shifts within each chapter. Without the extra space between paragraphs, she did not expect those shifts. The sudden appearance of a different character or setting knocked her right out of the story while she oriented herself.

In other words, the spaces between paragraphs are just as important to the flow and voice of your content as the rests between notes are to the sound and performance of a piece of music.

Here are two examples of how voice can affect the reading experience, even when paragraphs convey the same basic information.

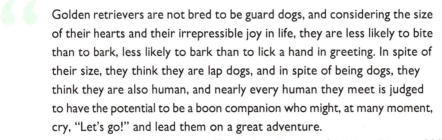

> Golden retrievers are not bred to be guard dogs, and considering the size of their hearts and their irrepressible joy in life, they are less likely to bite than to bark, less likely to bark than to lick a hand in greeting. In spite of their size, they think they are lap dogs, and in spite of being dogs, they think they are also human, and nearly every human they meet is judged to have the potential to be a boon companion who might, at many moment, cry, "Let's go!" and lead them on a great adventure.
> —Dean Koontz, from his novel *The Darkest Evening of the Year* (Koontz 2007).

> Goldens are outgoing, trustworthy, and eager-to-please family dogs, and relatively easy to train. They take a joyous and playful approach to life and maintain this puppyish behavior into adulthood. These energetic, powerful gundogs enjoy outdoor play. For a breed built to retrieve waterfowl for hours on end, swimming and fetching are natural pastimes.
> —American Kennel Club, *Golden Retriever* (American Kennel Club)

Dean Koontz writes two long sentences with deliberate word repetition, rhythmic phrasing, and intricate pauses. At a total of 97 words, this paragraph builds suspense as the reader keeps seeking its climax.

The American Kennel Club writes four short sentences. The first three use a basic subject-verb-object construction. Only the last sentence introduces sentence structure variety.

Both paragraphs describe golden retrievers accurately. Both paragraphs represent their brands well. Koontz can reuse that paragraph in his memoir, on his website, or in a letter to friends, and it will blend in perfectly with the rest of his writing. The Kennel Club can reuse its paragraph in a brochure or in the voiceover for an online video, and it will exactly represent the personality of the organization.

But if you put the Kennel Club's description into the middle of a Koontz essay or novel? The reader will be thrown right out of the emotional experience of reading and wonder what happened.

Your company's voice is crucial to building your brand, distinguishing your content from your competitors' content, and developing relationships with your customers. At best, content that strays from your voice standards provides a bit of friction for your customers. At worst, breaking voice provides an unpleasant experience and drives customers away.

Voice Standards in Your Company's Content

In a world of personalization at scale, inconsistencies in voice quickly become apparent to your customers:

- Content with different voices does not flow seamlessly from one paragraph to the next.
- Content with too much voice interrupts the message and distracts the reader.
- Content with no voice at all is boring and does not differentiate your content from other information that your customers encounter online every day.

Does your company have a single defined voice for all departments to work toward? Many companies think about voice only when they develop marketing strategies. And even when companies invest deeply in developing and documenting voice guidelines, those guidelines rarely make it from marketing to other content silos.

- Education and learning content teams might develop their own voice guidelines, which might not match up with marketing's guidelines.
- Customer support teams might literally use voice, as they prepare scripts for agents who work with customers over the phone or through chat. Sometimes these teams include voice guidelines for support articles. Automated chatbot content might also adhere to specific voice guidelines.

- Product documentation often tries to have no voice at all. This lack of definition often results in product documentation that is overly academic, unnecessarily wordy, or so formalized that customer comprehension suffers.

If voices across the teams within your company sound like they are from entirely different organizations, you are making your customers work harder than necessary. It isn't the customers' job to reconcile the differences in their experience with your company during a sales review versus a training session versus a support interaction.

How Voice Builds Personal Experiences

In personalization, we try to match content to the person who experiences it. But we also try to make our company a *person*: We need customers to feel that they are engaging with us as people, not as a faceless uncaring corporation or a weird stalker robot.

When we standardize content to enable reuse and provide consistent experiences for customers, we need to bring our voice into *all* our content. Otherwise, the customer journey suffers from unexpected zigs and zags.

And let's admit it: Customers do not follow the journeys we map out for them. They just don't. Rather, they use our content to tell themselves a story about what is going to happen and what they are going to do.

For example, we might expect customers to "meet us" by clicking on a link in social media. They might then explore the product page and read user reviews on the website. From there they purchase the product. They follow our instructions to set up the product. They refer to our support content if they run into trouble.

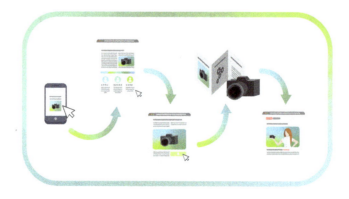

But in reality, a customer's journey often looks quite different.

Maybe they come to our website to look at documentation after a Google search for a product that performs a specific task. The first article they read is about troubleshooting the setup process, because that article has received the most traffic on our site and is therefore highlighted in our automated FAQ sidebar. The customer now has the idea that the setup process is difficult and fraught with peril, so they move on to the next item in their search. The next day, their Facebook feed shows an ad for our product. The customer thinks that's creepy but clicks through anyway because the offer is really good. Only then do they read the product marketing.

Eventually, the customer might make a purchase, at which time they receive full access to our product documentation and other resources available only post-purchase. But the path they took to get there was not the one we mapped out for them. Even though they made a purchase, if our content doesn't resonate with them at every point, in whatever order they experience those interactions, they might never come back.

All Content Needs a Voice

Marketers have long called for companies to develop their voices. The company voice connects with customers and encourages them to stay loyal over time. Even when your company produces physical objects like shoes, cars, or toys, your company's or products' voice sells those objects more than the objects themselves do.

Because, let's face it: There isn't much difference between a Nike running shoe and a New Balance running shoe. It's the companies' *voices* that differ. Nike lures customers who resonate with the ideas of an active lifestyle, celebrity athletes, and motivational taglines ("Just do it!"). New Balance is all about helping runners overcome various physical challenges, such as having wider-than-average feet or dysfunctional pronation. Nike is sexy; New Balance is steady.

Voice is arguably even more important when you're selling an idea, such as banking services or auto insurance, rather than a physical object. Making a shoe or toy look distinctly different from another isn't that difficult. But how do you help customers distinguish between checking accounts or insurance policies?

To show just how much impact voice can have, let's look at two examples from the world of auto insurance: Geico and State Farm.

A quick factual comparison of the two corporations might lead you to believe that there isn't much difference between the companies. They're about the same age, their product offerings are similar, and they're pretty close in industry ranking. The following table shows their bona fides:

	Geico[a]	State Farm[b]
Year founded	1936	1922
Founders	Lillian Goodwin, bookkeeper, and Leo Goodwin, insurance salesperson	George Jacob "G.J." Mecherle, retired farmer and insurance salesperson
First product	Auto insurance	Auto insurance
Product offerings	Insurance for: • Vehicle • Property • Life • Other	Insurance for: • Vehicle • Property • Life • Other Banking and investments
Industry rank[c]	#2, with a little over 13% of all auto policies in the U.S.	#1, with a little over 17% of all auto policies in the U.S.

[a] https://www.geico.com/about/corporate/history/

[b] https://www.statefarm.com/about-us/company-overview/company-profile/state-farm-story

[c] "The 50 Largest Insurance Companies" (Everquote 2019)

And yet, if you've watched a 21st-century Super Bowl (including the ads), you can probably immediately describe the very distinct personality of each company:

- **Geico:** fun-loving, quirky, friendly, silly, youthful
- **State Farm:** caring, neighborly, gentle, forgiving, experienced

You might remember some of the absurd Geico commercials in recent years:[1]

- The wedding reception, as the groom who was raised by wolves introduces wedding guests to the members of his wolf pack while the bride shows her bridesmaid how much money she saved with Geico
- The pig at the DMV, using his Geico mobile app to show his proof of insurance so that he can get a driver's license
- The new apartment dwellers who have a "clogging" problem (the family upstairs dancing all day and night) but save money with Geico insurance
- Cynthia, the HOA representative who chops down a mail box and planter because they don't meet HOA specifications, but the family saves money with Geico insurance

Geico's voice carries through from its television ads to its highway billboards to its website, social media, and even customer support.

And yet, if a Geico customer is in a car accident, it doesn't matter how much they appreciate the company's odd humor. They're worried, stressed, and probably wondering whether their insurance bill is going to go up. They're not in the mood for CGI pigs or tap-dancing neighbors. They just want to know if they're facing a higher bill or not.

This is the risk of an especially distinctive company voice: It's easy to overdo it.

Geico understands that challenge. Figure 8.1 shows an example from Geico support.

---

[1] Geico YouTube Channel [https://www.youtube.com/user/GEICO].

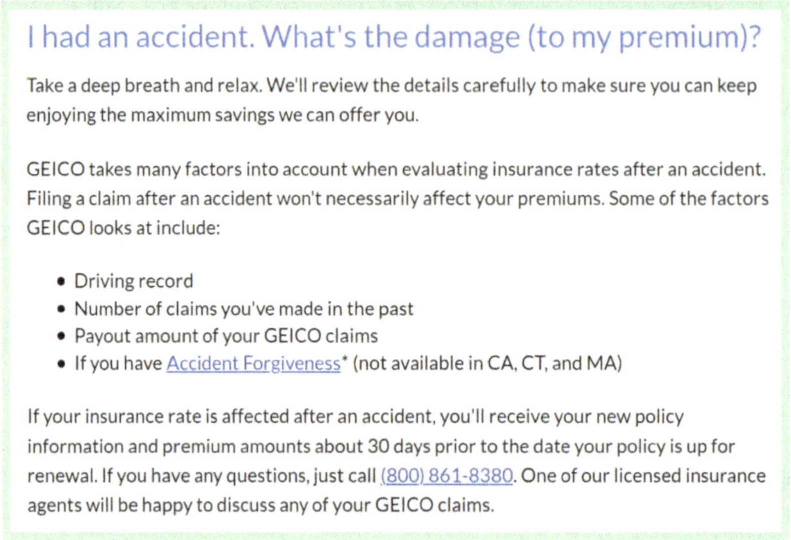

> ## I had an accident. What's the damage (to my premium)?
>
> Take a deep breath and relax. We'll review the details carefully to make sure you can keep enjoying the maximum savings we can offer you.
>
> GEICO takes many factors into account when evaluating insurance rates after an accident. Filing a claim after an accident won't necessarily affect your premiums. Some of the factors GEICO looks at include:
>
> - Driving record
> - Number of claims you've made in the past
> - Payout amount of your GEICO claims
> - If you have Accident Forgiveness* (not available in CA, CT, and MA)
>
> If your insurance rate is affected after an accident, you'll receive your new policy information and premium amounts about 30 days prior to the date your policy is up for renewal. If you have any questions, just call (800) 861-8380. One of our licensed insurance agents will be happy to discuss any of your GEICO claims.

Figure 8.1 – Geico support page[2]

Geico allows some of its trademark humor ("What's the damage to my premium?") and some friendly language ("Take a deep breath and relax") into this support content. But the company doesn't make customers wade through a lot of personality to get to the answer they want. The content provides that information in a clear, simple manner.

Note, too, that the restrictions, limitations, and hard-nosed business content contain *no* humor or fun. However, even that text conforms to the overall writing style of the rest of the article, using contractions and keeping the tone friendly.

In this simple example, we see that even a tiny bit of voice—when paired with strong standardization at the word and sentence levels—is enough to make the content *feel* like it belongs to Geico.

As an old adage notes: products don't sell products, people sell products. Now that people spend so much time online, always connected, continuously engaging with content, it's more important than ever for the full breadth of your company content to have a unified voice. Your company's *person* cannot begin and end with marketing. To provide personalized experiences at scale, every content creator needs to take part.

---

[2] Source: Geico.com [https://www.geico.com/claims/claimsprocess/accident-impact-on-rate/]

How to Standardize Paragraphs

The goal of standardizing paragraphs is not to make each one read exactly the same way. The goal is to ensure that every paragraph flows easily with those that come before and after it.

The steps to standardizing paragraphs include:

1. Define the voice level.
2. Define the reading level.
3. Define the paragraph parameters.
4. Configure your content optimization software or update your style guide.

[1] Define the Voice Level

Voice level defines how much or how little voice to include in a paragraph. Remember: It does not take much voice for a paragraph to become overloaded. And too much voice in every paragraph can make for difficult reading. You can get more reuse out of a given content component if you do not set every paragraph to maximum voice.

Suppose that Geico rebrands itself and defines its new voice as Pirate English. If the company adds maximum voice to each and every paragraph, here's how Figure 8.1 might read:

I had an accident. Wha's th' damage (t' me premium)?

Take a deep breath 'n relax. Filin' a claim won't necessarily affect yer premium. We'll review th' details carefully t' make sure ye can keep enjoyin' th' maximum savin's GEICO can offer ye. When evaluatin' yer rates aft an accident, GEICO loots many factors, includin':

- Drivin' log
- Number o' claims ye've made in th' past
- Payout amount o' yer GEICO claims
- If ye 'ave Accident Forgiveness (nah available in CA, CT, 'n MA)

If yer insurance rate be affected aft an accident, ye'll receive yer new policy information 'n premium amounts about 30 days prior t' th' date yer policy be up fer renewal. If ye 'ave any riddles, jus' call (800) 861-8380. One o' our licensed insurance agents will be happy t' discuss any o' yer GEICO claims.

Quite a slog to get through, right? Did you read the whole thing or skip and skim?

Instead, the company could use a lower level of voice and get something like this:

I had an accident. What's th' damage (t' me premium)?

Take a deep breath and relax. Filing a claim won't necessarily affect yer premium. We'll review the details carefully to make sure ye can keep enjoyin' the maximum savings GEICO can offer ye. When evaluating yer insurance rates after an accident, GEICO takes many factors into account, includin':

- Driving record
- Number o' claims ye've made in the past
- Payout amount of yer GEICO claims
- If ye have Accident Forgiveness (not available in CA, CT, and MA)

If yer insurance rate is affected after an accident, ye'll receive your new policy information and premium amounts about 30 days prior to the date your policy is up fer renewal. If ye have any questions, just call (800) 861-8380. One o' our licensed insurance agents will be happy to discuss any of yer GEICO claims.

The lower level of voice in the support article is much easier on the customer. There are enough touches to convey the company's voice, but not so many as to get in the way of comprehension.

But how do you define which level voice to put in different paragraphs?

If we draw a spectrum for voice across a general set of content silos, it might look like this:

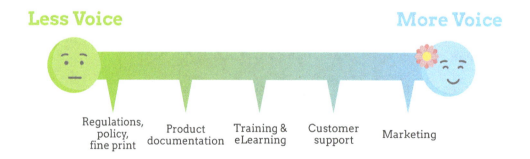

Less Voice   **More Voice**

Regulations, policy, fine print   Product documentation   Training & eLearning   Customer support   Marketing

On this spectrum, regulatory, policy, and fine-print content have the least voice. (Even Geico does not reference pigs getting driver's licenses in its regulatory content!) As we move from regulatory content, through product or training documentation, all the way to marketing content, the addition of more voice makes sense. The content at the right side of the spectrum is intended to be more engaging than the serious policy information on the left side.

Here's where things can get tricky. In a traditional publishing paradigm, writing to the appropriate level of voice is easy. We write a paragraph as part of a static document. And we know the purpose of that document in advance. We know the paragraph is seen there and nowhere else.

But in a world of personalization at scale, content creators don't know ahead of time where paragraphs will appear or in which output types customers will read them. In this new paradigm, we need to define voice levels based on the goal of each paragraph, not on the silo or output type or content order in which that paragraph appears.

In most company content, each paragraph typically has one of these goals:

- **Alert:** For example, detail an end-user licensing agreement or the conditions of a loan
- **Inform:** Describe a procedure or present product specs
- **Teach:** Train customers to implement a complex technology solution
- **Reassure:** Support customers who encounter issues
- **Engage:** Attract new customers or inspire customers to buy more

When you focus on the goal of each paragraph instead of on the output in which it appears, the voice-level spectrum looks more like this:

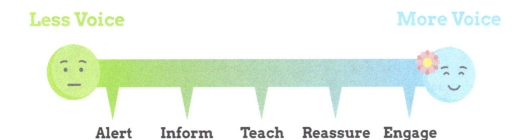

Less Voice   **More Voice**

Alert   Inform   Teach   Reassure   Engage

If your enterprise voice is quirky and fun, how much of it should you put into a troubleshooting procedure? If customers are in trouble, do they want a lot of fun? Or should your troubleshooting content tone it down a bit to respect a customer's likely emotional state (frustration, confusion, maybe even anger)?

In your paragraph standards, then, focus on the paragraph's purpose. Define the purposes for which authors can use a stronger voice or personality and those for which they should aim for a more basic, informative style. You can (and should) design a certain amount of flex into your paragraph standards to give the customer the best experience.

> Flex is an important framework when we're crafting voice. We flex our voice and tone to meet customers where they are in the experience. Our voice takes on different characteristics, or attributes, in the marketing pages than it expresses in support content. We consider the variation in customer needs throughout those very different touchpoints and adjust our voice accordingly.
> —John Caldwell, author of *Voice and Tone Strategy* (Caldwell 2020)

[2] Define a Target Reading Level

Reading level is an assessment of the level of formal education a customer is likely to need to understand a piece of content. Variations in reading level can wreck a customer experience in a heartbeat. You might have experienced this issue if you've ever followed a link from a support article about troubleshooting a desktop app to a system administration guide. Suddenly the content is a wall of text with long sentences and an academic writing style full of passive voice and jargon.

By standardizing on a target reading level, you can help ensure that all your content can be delivered to customers without requiring them to change their reading-level skills between components. Several reading-level scales can help you evaluate the complexity of your content. Some of the more common scales are:

- Flesch Reading Ease
- Flesch-Kincaid Grade Level
- Gunning Fog Index

Many of these formulas are available in software that can evaluate content as you write, providing immediate feedback for content creators. These tools use criteria such as number of syllables in a word, number of words in a sentence, or number of letters in a sentence to determine how readable content is.

Flesch Reading Ease

Rudolf Flesch invented the Flesch Reading Ease score in the 1940s. This assessment measures how complicated a passage of text is. The scale is a bit counter-intuitive, the lower the score, the more difficult the text. However, it's one of the most regularly used reading-level formulas for English content.

The Flesch Reading Ease formula looks at the total number of words, sentences, and syllables:[3]

$$206.835 - 1.015\left(\frac{\text{total words}}{\text{total sentences}}\right) - 84.6\left(\frac{\text{total syllables}}{\text{total words}}\right)$$

Flesch Reading Ease is then scored on a scale from 0 to 100.

Score	Grade Level	Notes
100–90	5th grade	Very easy to read
90–80	6th grade	Easy to read
80–70	7th grade	Fairly easy to read
70–60	8th and 9th grade	Easily understood by 13- to 15-year-old readers (A score of 65 is the level to shoot for using Plain English.)
60–50	10th to 12th grade	Fairly difficult to read
50–30	College	Difficult to read
30–0	College graduate	Very difficult to read

---

[3] *How to Write Plain English* (Flesch 1979)

Flesch-Kincaid Grade Level

In the 1970s, Rudolf Flesch teamed up with J. Peter Kincaid to develop the Flesch-Kincaid Grade Level readability test for the U.S. Navy. Rather than scoring text on a scale from 1 to 100, the Flesch-Kincaid Grade Level test uses the U.S. educational grade level as its scale. This makes it easier for people to understand and correlate the score with education level. This scale is also commonly used for English content.

The formula is similar to the Flesch Reading Ease score, but the weighting and result are different:[4]

$$0.39 \left(\frac{\text{total words}}{\text{total sentences}} \right) + 11.8 \left(\frac{\text{total syllables}}{\text{total words}} \right) - 15.59$$

The result of the Flesch Reading Ease score is a grade level based on the U.S. education system. For example, a score of 6.3 indicates that the text should be understandable by someone in the 6th grade.

Gunning Fog Index

The Gunning Fog Index aims to "eliminate fog" in business writing. The system aims especially at subject matter experts who write as part of their jobs but are not writers by profession. Developed in the 1940s, when calculation was done by hand, this formula is simpler than that of its contemporary, the Flesch Reading Ease system. The Gunning Fog formula is:[5]

$$0.4 \left[\left(\frac{\text{total words}}{\text{total sentences}} \right) + 100 \left(\frac{\text{complex words}}{\text{total words}} \right) \right]$$

Any word of three or more syllables is considered a complex word, with some leeway for suffixes. The resulting score indicates a grade level.

Writing to a Reading Level

Whichever reading-level scale you choose, you need to determine your target reading level based on the needs of your customers—not on the gut feel of your content teams. You also need to give content creators the tools to measure reading level so they can adjust their content accordingly.

---

[4] "Derivation of new readability formulas (Automated Readability Index, Fog Count and Flesch Reading Ease Formula) for Navy enlisted personnel" (Kincaid 1975)

[5] *The technique of clear writing* (Gunning 1968)

The average reading level in the U.S. for people who have English as a first language is somewhere between 6th grade and 8th grade.

Do not confuse reading level with intelligence. Smart people who will be thrilled to engage with your content might have lower reading-level scores in English than you might expect. Perhaps English is their second or third language. Or perhaps they prefer to learn from images or the spoken word than written text.

Regardless of which metric you use, the best advice is to keep your sentences short and your words simple. Customers who love to dive into Proust on the weekends still don't want to keep a dictionary nearby just to understand your company's content.

[3] Define Paragraph Parameters

Paragraph parameters include guidelines for paragraph length and presentation.

Length

We strongly recommend that you standardize on short paragraphs for most content. Two to three sentences is a good length for reading on a screen.

Shorter paragraphs are typically more readable. Readers can more easily scan a page to find the content they need when paragraphs are short and separate by a good amount of space. Short paragraphs are also easier to consume on mobile devices and through voice interfaces.

Short paragraphs are more versatile for personalizing content. They make it easy to separate common content from unique content. Common paragraphs can be reused, whereas unique content is provided only when and where needed.

You can allow longer paragraphs where appropriate for the content. Academic, scientific, and legal content might need to comply with external standards. You might be required to include a high volume of information in a single paragraph for regulated content, based on the regulatory body's requirements. As long as you define these situations in your content standards so that longer paragraphs are intentional, you can still reap the benefits of standardizing on shorter paragraphs in the rest of your content.

Presentation

Some content is clearer when presented in lists, notes, or tables than in regular paragraphs. Sometimes the decision to use alternatives to the paragraph must be left to the content creator. Other times, content clearly benefits from being consistently created as a list, note, or table.

For example, some of our customers define a *note* as being for supplemental information that is not necessary to the main content. Other customers define a *note* as being a way to highlight important content. Neither standard is inherently correct; it's each company's definition of and commitment to the role of a note that builds unity across their content.

[4] Configure Content Optimization Software or Update Your Style Guide

After developing your paragraph standards, you need to document them and enforce them.

The most effective way to do so is to configure them in your content optimization software. A less expensive (but also less effective) way is to document them in your style guide.

In either case, train content creators to use the specified resources to write to your voice standards.

Have You Lost Your Voice?

If you don't know whether your company has a defined voice and tone, check with marketing. Even without a formal strategy, your marketing writers likely have guidelines and examples that other content teams can start with.

If your company has a loosely defined brand voice but has not documented specific standards for how content creators should incorporate voice into content, you can start with the brand voice.

Summary

Different types of information require more or less voice. Take care: Overdoing voice is easy. So is underdoing voice and creating content so bland and boring that no customer can engage with it. Voice level, reading level, and paragraph length and presentation all have a role to play in developing paragraph standards that successfully convey the voice of your company.

CHAPTER 9
Managing Components

Components are the building blocks of personalized content. Words, sentences, and paragraphs all roll up into content components. Several components assemble to form a personalized experience. Each component is matched to the customer at the right time, on the right device, in their preferred language.

Like words and sentences, components must be standardized to flow together seamlessly in whatever order the customer encounters them.

What Is a Component?

A component is an independent unit of content that can stand on its own. It is a building block of content that can be used many times. *Component* might be a new term for you, but you've likely encountered the concept under a different name:

- Topic
- Module
- Block (or content block)
- Chunk
- Nugget
- Snippet

A customer might see the component as one section of a web page, in the context of other components. Or a customer might see the same component as a completely standalone piece, such as a banner ad on a website or an answer in a support chatbot. Components are self-contained units of information. Each component is an easily digestible piece of information.

What Makes a Component a Component?

A well-written, reusable component contains exactly the right amount of information, nothing more and nothing less. It makes sense all by itself, yet also fits seamlessly with other components. That's a lot for one little chunk of content to do.

To be effective, a component must have all the following qualities:

- Focus on a single subject
- Ability to stand alone
- Minimal or no formatting
- A business purpose
- Company-wide, rather than silo-ed, application

Let's look at each of these qualities in more detail.

Components Are Extremely Focused

The equation is fairly simple: The longer your component, the less likely it will be reusable. The opposite is equally true. The shorter and more focused your component, the more likely it will be reusable in a variety of settings.

A good rule of thumb is to limit the size of your component to one heading level. That is, do not put subheadings inside a component. If your content needs a new subheading, that's great; start a new component there. Components are more reusable when they contain one title and a focused body of content.

You can arrange components to create heading levels in the output. The same component can be a level-2 heading in one output and a level-3 heading in another, without any changes to the source content. (To quote one of our customers: "MIND BLOWN!")

A component needs to be small enough to contain one—and only one—thought, point, or idea, yet large enough to cover the entire idea. A component must not depend on other components to make sense. Each component must be self-contained. After all, you never know where a single component will be reused, so it should contain everything your reader needs to understand that single idea and not one word more.

> It's kind of like a gourmet, make-at-home meal kit. Components, like meal kits, stand alone. Everything you need to create the meal is contained within the box delivered to your door. The ingredients are like the sentences in the component. They are all necessary and work together to create a single meal (or describe an idea). Also, you don't need any other meal kits to understand the meal kit you have. It functions and makes sense by itself.
>
> —Jonathan Chandler, Technical Publications Manager

Components Must Stand Alone

For components to work in a personalized experience, they must make sense all on their own. Reusable content cannot rely on information in other components that come before or after it. You never know where a particular component will be used, so each must stand on its own.

Three best practices for standalone components include:

- Avoid dependent language
- Use links sparingly
- Start strong

Avoid Dependent Language

Specific content is more difficult to reuse. For example, if you include specific product names in a feature description, you can't use that content for other products—even if they include the same feature.

Table 9.1 contains some examples of the same information, using dependent language and corrected to use independent language.

Table 9.1 – Examples of dependent and independent language

Dependent language	Independent language
Now, you can make video calls from your wrist.	You can make video calls from your wrist.
Previously in this document, Example A showed that puppies are cute.	Puppies are cute. See Example A.
See Example B **later in this document.**	See Example B.
As discussed before, kittens are also cute.	Kittens are cute.
Currently, the app does not offer an alternative.	The app does not offer an alternative.
In this release, kittens and puppies are included.	Release 123 includes kittens and puppies.

Build independent language guidelines into your standards for words, sentences, and paragraphs.

Use Links Sparingly

Inline links and cross-references are another type of dependent language. When you write in a non-reusable, monolithic paradigm, including links and cross-references to content that exists elsewhere in the deliverable is easy.

When you write for reuse, including links and cross-references to other components *prevents* the component from being reusable. Authors can't always know whether the cross-referenced component will be included in the same output. The closer you get to delivering personalized content, the higher the risk that inline links and cross-references will break.

The good news is, when content is chunked well and tagged consistently, the need for links and cross-references is greatly reduced. Customers are much more likely to find exactly the information they need quickly. They don't need to scan through a long document or rely on authors to send them to other chapters just to find the one paragraph that answers their question.

Make sure that all the information you need for a component is contained in that component. If other information is critical to understand the component, include that information. And resist the urge to link or cross-reference any "nice to have but not critical" content.

Start Strong

Pretend that customers will see only the title and first sentence of your component. Make sure that first sentence answers the question that the title implies.

Often, we like to wax poetic and ease in to the topic of our component. Doing so forces customers to read (or worse, listen to) extraneous words before they get to the *real* content. Easing into a topic reduces overall usability and increases the risk that the customer will leave before they realize that the answer they need is just a few sentences away.

Here is an example of a chunk that starts strong.

(source: Wikipedia[1])

Even if you read only the first sentence, you can answer the question, "Who was Komatsu Imai?"

---

[1] https://en.wikipedia.org/wiki/Komatsu_Imai

It's going to take work because the content creator's intent has got to be to look at those chunks and decide, "What am I going to sub-containerize within this thing?" There are no heuristics that I know of that could ever computationally dictate or figure out that intent. It has to be the creator deciding, "This is a logical unit, different from a hierarchically structured content unit."

—Mike Iantosca, Enterprise Content Strategist

Components Have Little or No Formatting

Components are created without much formatting. Content creators do not apply heading styles, add fonts, or rearrange margins. They do not dictate what the content will look like. Instead, they focus on creating the content itself.

Design elements are added at the end, during publishing. This arrangement allows components to flow into any output type for any channel or device. The published output always has the right fonts, the right colors, and the right design, because you already created a design for the content to flow into. That design is separate from the content itself.

Many authoring tools enable content creators to preview the content in one or more output formats. The preview option enables content creators to visualize how the content will look when all design elements are applied. This preview can help creators transition from the old ways of writing, where so much of their time was spent formatting, to new ways of writing, in which they can focus on the content without the manual steps of formatting it.

One technical publications manager we know says, "I feel like the need to preview and see how the content will look in the final form becomes less of a habit as writers get more comfortable with structure and outputs become very stable. I think this trust that the content will 'look right' can help reduce content creation time over the long run, as writers do less publishing and checking throughout the project."

We often see components described as being format-free. That's mostly true. In some cases, content creators need to add *some* formatting, such as table borders or emphasis. Your content standards should define which types of formatting to add and when to add them. That's how components can be reused in multiple scenarios and still provide a cohesive customer experience.

Components Serve a Business Purpose

Each component must serve a business purpose. If you cannot point to a specific business reason for a component to exist, then why invest resources in creating and maintaining it?

Content serves all kinds of business purposes. Here are some common ones:

- Engage a person who has never heard of your company.
- Convince a customer to buy additional products.
- Instruct a customer about how to use a product feature.
- Teach a customer how to use your product to achieve their job goals.
- Help a new customer make a purchasing decision.
- Answer a customer's question.
- Solve a customer's technical issue.
- Confirm a transaction.
- Establish a reputation for thought leadership or specialized expertise.
- Onboard, develop, and retain talent.

Typically, a component serves only one of these purposes. Many components come together to serve many purposes.

Think about it. If you are trying to engage a stranger, you're unlikely to do so by giving them detailed instructions for how to use a product.

On the other hand, if an existing customer needs help using a product, you don't want to clutter the step-by-step instructions with verbiage about how versatile the product is or list all the other things it can do. The customer is already engaged, has already bought the product, and is asking for help. Your component must provide that help without forcing the customer to wade through a bunch of words they don't need.

If your component describes a new feature and provides instructions for how to use it, then it's doing too much. You can easily solve this issue by splitting the content into separate components. Now, you have both a feature description and instructions that you can use whenever and wherever you need.

Likewise, if you want to serve two business purposes in a single email—say, confirm a transaction and establish a reputation for thought leadership—you can combine two single-purpose components. Use one that confirms the transaction ("Your order has shipped") and one that establishes your company's reputation ("Our CEO wrote this blog article; we think you'll enjoy it").

Components Speak for the Business, Not for a Silo

A component must reflect your business, not the organizational silo that created it (or the individual content creator who wrote it). Trust us: Your customers do not care who created the content. What matters is that they receive the information they need when they need it.

You might assign organizational silos to create and manage certain types of components. But even if systems are not yet in place to facilitate company-wide reuse, your components should be created as if every content team in the company will use them. Because someday, they will.

To make sure each component reflects your business, focus on standardizing your terms, sentence structures, and voice. And then enforce those standards. This is the only way to mix and match a wide variety of components and the only way to provide personalized experiences at scale.

Transforming Legacy Content to Components

Your organization likely has legacy content locked away in monolithic PDFs. You know the ones: linear documents with hundreds of pages. Navigation in the form of a table of contents (TOC)—several pages in its own right—and possibly an index (another dozen or more pages). Running heads, cross-references, mini-TOCs within each chapter, and other navigation help. Ultimately, we're talking long linear documents with hundreds of pages … or more.

The longest PDF Regina has encountered since she started tracking was a data dictionary of about 12,000 pages. (If you can beat that, you have our deepest sympathies. Call us. We can help.)

Even today, thirty-plus years since topic-based content became a thing, we find that PDFs of up to 800 pages are not uncommon. One technical writing team at a large financial services company burst into laughter when Regina mentioned a project in which she helped authors componentize an 800-page manual into a library of standardized, reusable content components. "Oh, 800 pages?" they said, "That's cute." And they sent her a 1,300-page reference guide.

Customers are going to find it difficult—nigh impossible—to find an answer within a document like that. How can you blame them if they give up and call support instead?

Legacy content also remains tantalizingly out of reach for your content creators. Source files are likely in Microsoft Word, Adobe Framemaker, MadCap Flare, Adobe InDesign, and so on. Or worse (but common, especially for older documents), nobody knows *where* the source files are. Even if your company is super-organized and you can find every file, successfully using components requires more than just chopping legacy content into chunks at every subheading.

> We started on a path to de-linearize content long ago with the advent of topic-based content models, and the trend towards more and more de-linearization continues. I take the view that a certain amount of non-linear content can and must co-exist within linear content; it isn't an all-or-nothing proposition. Both prescribed assemblies and prescribed navigation will co-exist with dynamic content assembly and delivery.
>
> —Mike Iantosca, Enterprise Content Strategist

How to Standardize Components

The steps to standardizing components include:

1. Define information types.
2. Develop component models.
3. Configure tools.
4. Follow component best practices.

[1] Define Information Types

Much of the information your company creates can be divided into three basic information types:

- **Conceptual** information describes an idea.
- **Instructional** information tells a person how to do something.
- **Support** information provides references and details.

Your content will have many, more specific information types. However, we find that simply considering these three basic types helps content teams begin to see where their own content can be separated into components.

Conceptual Components

Conceptual components **describe**. This information type is all about conveying an idea. Conceptual components include content such as:

- Descriptions of products, features, or services
- Value propositions
- Overviews, summaries, and introductions
- The "hook" of a white paper or solution brief
- A policy statement
- A question-and-answer pair (such as for an FAQ list or chatbot)

Most marketing content is conceptual. After all, such content exists to convey the idea that the customer has a need and that buying the marketed product or service will fulfill this need. Marketing content needs to tell compelling stories that resonate with customers without burying (or boring) those customers with a lot of instruction or excruciating detail.

A significant portion of training content is also conceptual. Product documentation and customer support also create a lot of conceptual components. Any time the content describes a *what* or a *why*, you are working with conceptual information.

Conceptual components are highly reusable.

For example, a feature description is a feature description. It doesn't matter whether the customer reads the description as part of your marketing efforts or as part of a post-purchase quick-start guide. You should reuse the same *Feature Description* component anywhere you need to describe the feature.

Not only is it wasteful to pay for two separately developed feature descriptions, it literally sends mixed messages to customers who encounter two different versions of the same content.

Let's build on that with another conceptual component: a *Feature Benefits* component. This component's job is to convey the benefit of using (buying) that feature.

What if benefits differ depending on customer behavior, region, season of the year, and so on? In this case, marketers might need to create several components, each of which describes a different benefit. Within the various customer experiences around this feature, each of those individual components can lead to the same *Feature Description* component.

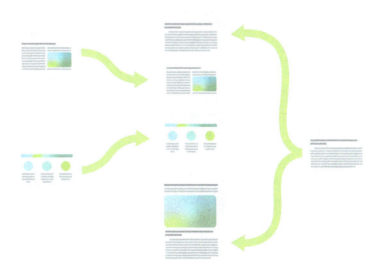

This is just one example of why identifying information types and keeping components separate are so important. You can reuse the same feature description in any experience that needs to describe the feature, keeping your messaging consistent and providing the most clarity for the customer. Then, you can provide whichever benefit components will be most compelling to different customers.

Simply defining *conceptual* as an information type isn't enough, though. You need to develop standards for different types of conceptual components that are specific to your content. Although most companies do have conceptual components such as *Overview* and *Product Description*, the standards for each of those components are unique to each company.

Beyond those common types are the truly unique component types. One of our customers defined a conceptual component called *Announcement*. Another defined a type called *Reason to Believe*. The important thing is that all content is created to the standards you specify.

Instructional Components

Instructional components describe how to **do something**. This information type is all about helping your customer achieve a goal. Instructional components include content such as:

- How-to article
- Installation procedure
- Lab exercise
- Troubleshooting tree
- Installation wizard
- eLearning activity instructions
- Video tutorial
- Walkthrough pop-up bubbles guiding a procedure

Instructions typically have high reuse across product documentation, training, and support. Although marketing might not provide much instruction, remember that people often explore instructional documentation before committing to a purchase, especially in a B2B context. Every component has a chance to make or break a customer relationship or a sale. (No pressure!)

Simply defining an *Instruction* component isn't enough. You need to develop standards for different types of instructional components. One of our customers defined several types of instructional components for their product documentation:

- *Simple* components for short, frequently reused procedures common to many products, such as how to verify the customer's version of Windows
- *Complex* components for long, complicated procedures that can branch depending on selections the customer makes earlier in the process

- *Troubleshooting* components for high-level procedures that point to *Simple* or *Complex* procedures for each step in the troubleshooting process

- *Training* components for procedure overviews that can be reused in instructor-led sessions and in eLearning modules

Support Components

Support components provide **detail**. This information type is all about providing details that people aren't going to memorize. Support components include content such as:

- Product specifications
- Measurement tables
- Product compatibility matrices
- Size conversion charts
- Research, survey, or study data
- Experiment results
- Machine requirements
- API methods across REST, SOAP, and JSON

Support components have great potential to improve the customer experience, if you standardize and componentize the information well.

For example, think about the last time you tried to find information about potential side effects in the pharmaceutical label that came with your medication. That label was probably printed on semi-translucent paper in about 5-point font, folded up 20 times, and taped to the side of the bottle. You needed to do reverse origami, find the beginning, and then skim through dense pages of text to find the side-effects section. Perhaps the information was presented in a table, showing the side effects by age, severity, or other factors.

Instead, imagine opening your medical provider's app on your mobile device or voice assistant to ask about side effects. Because you are securely signed in to your medical record, which includes your prescriptions, you can immediately access specific information about side effects that match your situation. If you want to know about other side effects, you expand your search to that specific information back. What you don't need to do is comb through hundreds of words that are totally irrelevant to the experience you are trying to have.

Simply defining a *Support* component for every information type isn't enough. You need to develop standards for different types of support components. For example, one of our customers makes silicon chips for use in electronics. The company's support component types include a *Pin Functions* component, which provides extremely detailed information about the chip hardware. Another customer defined a *Belt Data* support component to provide technical and mechanical specifications about conveyor belts.

The more your content can comply with standards, the more easily your content creators and your dynamic delivery systems can make the right choices about which components to show to which customers.

These three information types are only the beginning. To prepare your content to deliver personalized experiences, you need to standardize at a much more detailed level. That's where component models come in.

[2] Develop Component Models

A component model defines standards for specific types of components. The model defines standards such as:

- The type of component
- What order to put the content in
- What content is mandatory and what content is optional
- Authoring guidelines for each element of content
- Boilerplate text, where used

To standardize components for personalized delivery, you need to model specific information types for your enterprise. How strict or how flexible your standards are depends greatly on the type of component you model.

Table 9.3 and Table 9.3 show component models for a feature description and a procedure.

Table 9.2 – Simple conceptual component model for a feature description

Part	Medium	Usage	Authoring Guidelines
Headline	Paragraph	Required	Focus on the benefit of the feature.
Feature	Paragraph	Required	Describe the product feature. Aim for 1–2 sentences, or about 25 words.
Visual	Image	Optional	OK to use photo, screenshot, or illustration.

Table 9.3 – Simple instructional component model for a procedure

Part	Medium	Usage	Authoring Guidelines
Title	Paragraph	Required	Begin with imperative verb.
Goal	Paragraph	Required	Describe the procedure goal in 1–2 sentences.
Prerequisites	Paragraph, list	Optional	Describe what MUST be done before starting this task.
System Requirements	List	Optional	List any resources required before starting. These are prerequisites to completing this task right now.
Role	Paragraph	Optional	Describe the role that performs this task.
More Info	Paragraph	Optional	Provide more information about the task.
Step Introduction	Paragraph	Required	Start with an infinitive verb: "To…".
Steps	Paragraph, image	Required	Use as many steps as necessary to take the customer through the procedure.

You might think that developing standards for instructional components is easy. After all, an instruction is an instruction is an instruction, right? It isn't as nebulous as a conceptual component or as detailed as a support component.

But here are some things you need to decide about your instructional content:

- Do you require an introductory sentence or paragraph? If so, what information should this introduction include?
- How many levels of sub-steps do you allow? (Hint: Customers have a hard time following more than one level. If you end up needing more, then you're likely trying to cram two or more tasks into one instructional component.)
- Do you allow notes, warnings, or "for more information" statements within the instructions? If so, how many and which kind?
- How do you indicate a potential risk or hazard to humans, data, or equipment?
- How do you include images within your instructional components: only one at the top, or within each step?
- Do you allow videos within an instructional component?
- What is the maximum number of steps that a single instructional component can include? (Hint: Best practice is to limit procedures to a maximum of 11 steps. If you need more, you're likely trying to cram two or more procedures into a single instructional component.)

Table 9.4 – Simple support component model for a product data matrix

Part	Medium	Usage	Authoring Guidelines
Title	Paragraph	Required	Use the product name followed by the product model: **Conveyor Belt 12345**
Table	Container	Required	Structure product data as a table.
Pitch	Number	Required	
Minimum Width	Number	Required	
Maximum Width	Number	Required	
Width Increments	Number	Required	
Molded Widths	Number	Required	
Hinge Style	Paragraph	Required	Reuse from warehouse topic.
Drive Method	Paragraph	Required	Reuse from warehouse topic.

Your component models will be more or less complex than these examples, depending on your content. We sometimes create very technical models, complete with XML tags or system field names, to help with configuration for a component content management system (CCMS). We've helped our customers develop calculation rules to automatically pull data from another system into the content. These technical models are useful for system configuration and automation.

Other times, we create simple models that focus only on the component's content and authoring guidance. These simple models help teams standardize components even before they move to a structured content ecosystem. Those simple models can be enhanced in the future to include technical information for system configuration. In the meantime, they are effective in helping improve content right away.

[3] Start Writing Components

You do not need to wait until you have a new CCMS to start changing how you create content. The best practices for writing standalone components for reuse also improve the quality of traditional content, whether the components are reused or not.

> A few years ago, we worked with a customer who was deploying a CCMS and structured authoring tool. When it came time to discuss changing and rewriting the content, the customer was completely surprised. He said, "I thought that the tools would automatically take care of that stuff." This is a well-educated person who has been in the technology field for many years. He did not understand that the organization of the content is not the same thing as the content itself. And organization, by itself, does not automatically make content reusable.

Writing chunks of content can seem limiting, particularly in the beginning. Thinking everything in your content is unique, and therefore cannot be reused, defeats the purpose of personalizing content at scale. Unfortunately, content creators often claim that this is the case. We usually find this claim to be largely untrue.

You need to find your reusable content happy place: The spot where your content is universal enough to be reused in a variety of outputs, but detailed enough to say something meaningful. It takes time and effort to learn how to write this way.

Writing for reuse takes a degree of vigilance. As new products are developed, product features or other new content required to support those products may seem unique or distinct. It takes vigilance to find how that new content fits into existing information types and component types. Perhaps it truly is unique content, but you should come to that conclusion only after attempting to fit it into the models you've developed.
—Jonathan Chandler, Technical Publications Manager

[4] Configure Systems

Your delivery systems and content creation tools should make it as easy as possible for your content creators to adhere to your standards.

Configure templates for each component type, so that content creators are never stuck staring at a blank page. Include any boilerplate text that applies. Put authoring guidance directly into the templates so that creators know what goes where. Build in reuse wherever you can, so that they don't waste time re-creating content or searching for reusable content they don't know exists.

Templates can simultaneously enforce standards while also hiding any technical complexity from content creators.

Summary

Components are standardized content assets, developed according to your content standards and created with reuse in mind. Creating component-based content is more than just chopping legacy content apart at each heading and calling each section a component. To be successful, you must have a strategy and standards that everyone follows.

CHAPTER 10
Reuse: The Only Way to Scale

To deliver personalized experiences at scale, your delivery platform needs a repository of modular, reusable content to draw from. The content must be current, accurate, and standardized so that it can mix and match wherever needed.

Building and maintaining this repository is not something you can do in a week. Content reuse is not something you can start on an ad-hoc basis. You need up-front planning to determine which content to reuse, how to reuse it, and how to manage it within and across organizations.

One of the biggest mistakes a company can make is to reuse content in an ad-hoc manner. Expecting reuse to "just happen" without a strategy usually leads to one of the following results:

- Writers continue to create redundant content because they are too hesitant, resistant, or overloaded to learn how to reuse.
- Writers get overly excited about all the reuse mechanisms provided by their software. They point to metrics on how much content they are reusing. They start turning themselves inside out to find reuse, referencing in a sentence here, a paragraph here, a word there. Then someone else starts reusing those pieces of content. Someone else throws a variable into the mix. Another team begins to conditionalize the text … and suddenly you have a knot of content with references your systems can't resolve.

The most effective way to *leverage* content reuse is to *plan* for reuse.

What Is Content Reuse?

Content reuse is the practice of creating one piece of content and then using that piece of content everywhere you need it. When you reuse content, each piece is created, translated, updated, managed, stored, and retired one, and only one, time.

Content reuse is the opposite of *copy, paste, tweak (CPT)*. When you use CPT, you might think you are reusing content. After all, you are starting with the same text. But, as soon as you copy a piece of content, you have already strayed far from reuse. And as soon as you tweak that copy, you are working with another version of the same content.

When people CPT the same content over and over again, they end up making so many tweaks in so many versions, that managing all those redundant versions becomes a real nightmare.

The Problem with CPT

Suppose an appliance manufacturer has two models of the same washing machine. One model's door hinges on the left side; the second model's door hinges on the right. Other than the door hinge, the two washing machines are identical.

As a content creator, you are asked to create two service manuals, one for each model. You decide to use CPT.

First, you create the service manual for the model with the left-hinged door. Then, to create the service manual for the right-hinged model, you paste that content into a new document. Next, you tweak any text and artwork that refer to the door hinge to reflect a right-hinged door.

As a result, you now have two almost (but not quite) identical versions of the same manual.

But then a reviewer notices a tiny mistake in the left-hinged service manual: a typo in one sentence. Because you copied that text, the mistake appears in both manuals. You are busy finalizing the left-hinged manual and promise yourself that you will fix the tiny mistake in the right-hinged version as soon as you finish your current task.

Time goes by. You enter review comments. You get busy. Things happen. And you forget to correct the right-hinged manual.

You now have two manuals that *should* be identical in all ways except the door hinge ... but they aren't.

The more you work on one manual or the other, the more the two books get out of sync. More models are added, each with a left- and right-hinged version. You continue to CPT from model to model, hinge to hinge.

After a while, you have so much content drift, you can hardly remember which service manual has which version of which text. Toss in white-labeled versions for other manufacturers that put their names on the machines, and before you know it, there are 40 versions of CPT manual. Then you're asked to send the manual for translation into 8 ... 12 ... 24 languages.

Which version do you use? Which is accurate? It's impossible to tell.

Content Reuse at Work

Now, take a breath and look at the same example, this time use content reuse rather than CPT.

Instead of writing the manual as a single file, you break it into components. Each feature is its own component. You write content about identical features one—and only one—time. You then use those content components, as is, for both models.

Whenever something differs between models, like the door hinge, you create multiple components, one for each version of that differing information *only*. To publish the manual, you build a map that contains each component you need, in the order you need them.

The majority of the two manuals contain the same components, written once and reused in both manuals. For sections that refer to the door hinge, the manual for the right-hinged machine uses the right-hinged content component, and the manual for the left-hinged machine uses the left-hinged content component.

No matter how many new features, new models, or white-labeled versions your company adds, you always have one, and only one, component for each feature that is identical across all models. You *never* have two components that contain the same information but written in different ways. And you never need to ask yourself which version is up to date and accurate.

Reusing content at the component level is just one part of a successful reuse strategy. We explore other methods later in this chapter.

Content Reuse at Scale

Content reuse is how you create a multitude of personalized experiences from a repository of standardized content.

When we say *at scale* (and we say that a lot), we aren't just talking about scaling content delivery. Content reuse also helps companies scale content development, management, and localization.

We don't mean to brag (okay, you caught us—yes we do), but one of our customers found that by standardizing in all five dimensions (output types, words, sentences, paragraphs, and components), they were able to reuse about 90 percent of the content in their equipment manuals.

The results were phenomenal.

They reduced time to market to just **10 to 15 percent** of their previous timelines. They eliminated the time and costs of desktop publishing in both the English source and translated versions. Their localization costs dropped to about **5 percent** of what they had been spending.

To deliver personalized experience at scale, you *must* reuse the same components of information repeatedly, in different output types and on different devices. **You cannot use copy, paste, tweak to deliver personalization at scale.**

How Standardization Facilitates Reuse

When content is standardized in all five dimensions, content reuse becomes a powerful engine for quickly delivering all kinds of outputs and experiences.

Let's review how standards facilitate reuse:

- **Output types:** The process of standardizing output types includes identifying opportunities to reuse content. You can then configure systems (and train content creators) to provide the right reused content component at the right time.
- **Components:** Component standards identify where to reuse content. Writing components according to reuse best practices ensures that they can be mixed and matched in different orders or even viewed on their own in the output or experience you deliver.
- **Paragraphs:** Paragraph standards help to ensure that the components read as if they are written by one content creator and delivered by one company. This unity is important to reuse content successfully.
- **Sentences:** Sentence standards protect customers from shifts in grammar and style. Customers don't need to do the mental work of adjusting for regional spelling, idioms, or even basic English conventions when experiencing reused components.
- **Words:** Terminology standards protect customers from needing to figure out whether a shift in terms indicates a shift in meaning.

It is technologically possible to reuse content without standardizing it. You can manually reuse content when you assemble an output. Systems can automatically deliver content to the customer at any point in the experience.

However, without standardizing all five dimensions, content reuse can produce a terribly disjointed experience. In the end, customers might lose trust in your content—and in your company.

What Is a Reuse Strategy?

You get the most ROI from content reuse when you approach it from a strategic perspective. A reuse strategy is a plan for which content you will reuse, how you will reuse it, and where you will reuse it.

Your reuse strategy identifies things such as:

- Which content is reused automatically
- Which content is reused manually
- What level of content can be reused
- Which mechanisms supply the reuse
- How reused content is managed

One of our high-tech customers hired us to help develop component reuse strategies for their product web pages. They had gotten themselves into a bind. They had begun to reuse components without first having a strategy.

Content creators were enthusiastically automating component-based web pages in the company's content delivery system. They were even more delighted by the ability to set up dynamic delivery that assembled pages on the fly as customers journeyed through the content.

Unfortunately, the ease of these new component-based approaches backfired. Soon, the company website had thousands of web pages made up of many more thousands of components. The site sent customers around in circles or dynamically serving up "related" components that repeated information the customers had already encountered. Customers were lost and frustrated in their quest to find information. And the content team was buried in components and pages to manage.

The company needed a reuse strategy to clearly identify:

- Which components to reuse automatically in a dynamic web page assembled on the fly to personalize the experience for the customer
- Which components to reuse as part of a static web page that was the same for every customer
- Where to reuse components on the web pages (e.g., anywhere, only at the top of the page, only at the bottom of the page)
- When to reuse components dynamically (based on customer data)
- Which components to reuse in follow-up email campaigns

Only after the company developed their reuse strategy were they able to provide a (much improved) personalized experience for each customer. They weren't able to solve the problem overnight. But everyone breathed a sigh of relief after they mapped out a dedicated plan for reuse.

When your reuse strategy is in place, your content standards implemented, and your tools configured, your ability to deliver personalized experiences at scale skyrockets.

How to Create a Reuse Strategy

A reuse strategy is a plan for what content you reuse, how you reuse it, and how you manage it throughout its lifecycle. This strategy defines:

1. **Granularity:** What is the smallest piece of content that can be reused?
2. **Matrix:** Where will the content be reused?
3. **Management:** Where will you store reused content?
4. **Governance:** Who can create, modify, or retire reused content?

[1] Define Granularity

Granularity defines the smallest unit of content that can be reused.

You can reuse content as small as a single character or as large as an entire document. The question, of course, is … should you? And if so, when?

The smaller the granularity, the more reusable the content—and the more attention needed to manage it. The larger the granularity, the less reusable the content.

Your reuse strategy defines which types of content to reuse at each of the following levels of granularity:

- Section
- Component
- Element

Section Reuse

Section reuse is the practice of reusing an assembly of content in its entirety. An assembly of content usually contains several components. For example, you might reuse a two-minute introduction section within every eLearning course for a product line. Or you might develop a library of case studies that can be reused in solution briefs and product web pages.

With section reuse, you create and assemble the content one time. You can then lock that assembly so that nobody can add or remove content. Content creators can bring the entire section into their outputs without changing any content within the section. Systems can automatically include the section when content creators make a new output based on a standardized output type.

Typical examples of section reuse in PDFs and print include:

- About Us, such as corporate history or executive bios
- Front matter, such as copyright statements and legal disclaimers
- Back matter, such as glossaries and indexes
- Introduction, such as "How to use this workbook"
- Appendix, such as error code definitions or troubleshooting guides

In a personalized experience delivered online, section reuse manifests as a series of components delivered together, in the same order, whenever the delivery platform provides that content.

One of our pharmaceutical customers has made great strides toward delivering more personalized experiences on their websites. Their reuse strategy includes section reuse to ensure that any claims they make about their medicines are accompanied by several supporting statements. A customer should never see a product claim (e.g., "This drug reduces heartburn") without accompanying data (e.g., "85 percent of patients reported heartburn relief after taking this medicine daily for 6 months").

Associating a product claim with supporting information is not just a good idea, it's the law. That's the reason why *Eating Well* magazine (as an example) devotes several pages to drug advertisements between the recipes. Life sciences companies cannot just provide marketing. They must provide a certain amount of data, cautions, warnings, and disclaimers, even in print.

The FDA also provides rules around how pharma companies must deliver this content in a digital platform. The marketing team Regina worked with cannot hide the supporting information behind a hyperlink. The content delivery must meet regulatory requirements for visibility, ease of access, and direct association between claim and support.

Other health authorities around the world also provide rules and guidelines around what content life sciences companies can and cannot show. Some rules depend on whether the customer is a patient, a physician, or simply a person exploring the website. Some depend on whether the region allows direct advertising to customers or whether content can be shown only to licensed health professionals.

The personalization strategy must account for factors such as region, customer type, and regulations. The reuse strategy must support the connections between elements, components, and sections to ensure that the right content is delivered to the right person on the right device at the right time—and with all its legally required components in their legally required order.

Component Reuse

Component reuse occurs at the level of individual chunks of content. A component is an independent unit of content that can stand on its own. It is a building block of content that can be used many times. Because it can stand on its own, it can be reused in many contexts without any change to the content.

Component reuse is the most common level of granularity for content reuse. With component reuse, individual components are reused exactly as they are. This level of reuse is why it is so important to define and create standards for each of your component types.

Every reuse strategy we've ever worked on includes component reuse. Not all companies reuse sections or elements. But everyone we've worked with, without fail, reuses components.

Component reuse is the main type of reuse for personalized experiences. Much of this book is dedicated to showing you how to make your content into reusable components.

Element Reuse

Element reuse happens at the paragraph, list, table, or sentence level. This level of reuse is often called *fragment* or *snippet* reuse.

Element reuse requires a little more planning and configuration than component or section reuse. You need to identify which elements can be reused and develop a management process for them.

Common examples of element reuse include:

- Warnings, cautions, tips, and other note fragments that appear throughout the content
- Tag lines, slogans, and other key phrases that need to be identical everywhere they appear
- Standardized sentences, such as the ones used in transactional emails ("Thanks for buying our stuff!") and chatbots ("Hello, how can I help you today?")

Many organizations use *warehouse components* to store and organize reusable elements. A warehouse component contains a collection of individual paragraphs, sentences, lists, or other reusable elements. You never deliver the warehouse component to a customer. Rather, content creators reference the reusable elements from the warehouse component to use those elements inside other components.

For example, one year, Regina worked with three equipment companies to develop reuse strategies. One of the companies makes manufacturing equipment for other manufacturers. The other two make medical devices.

All three companies are required to include certain warning statements and symbols on their equipment and in their documentation. All three had legacy content in which various symbols were included next to various warnings. Sometimes, the same symbol was used with different warnings. Sometimes, the same warning was used with different symbols. And sometimes, the same warning was written differently in different contexts. No one-to-one relationship of symbol to warning or standardized warning text existed.

Writing content in different ways can lead to confusion for customers. In this context, confusion can cause equipment damage or bodily harm. Using the same symbol to mean different things or different symbols to mean different things can be dangerous.

As each team worked to standardize their written content, they also collected examples of inconsistencies in symbol usage. Each team planned to reach a future state of one standardized symbol per one standardized warning.

But as Regina developed reuse strategies with each team, it became clear that some symbol-and-warning or even warning-only components simply could not be standardized.

These equipment manufacturers (and many others) can't simply swap out images and update text in their documentation. Many symbols and warnings are required on the equipment itself: molded into plastic, etched into metal, or adhered as industrial-strength stickers. Documentation must replicate those symbols and warnings exactly. So, if nobody upstream is committed to consistency, then downstream teams in marketing, documentation, training, and support must follow suit.

As the teams standardized each symbol and warning that they could, they added those elements to a warehouse component. Content creators were at least able to reuse those elements, helping to ensure consistency and eliminating the need for certain warnings to be re-reviewed or re-approved after initial approval.

As those reuse libraries grow, the companies will start to reap the benefits of standardized content. They then hope to convince content creators in the product planning and development stages to apply these content standards from the beginning, before content is etched into their equipment.

[2] Develop a Reuse Matrix

To identify your reuse opportunities at each granularity level, analyze your content and create a reuse matrix (see Table 10.1).

Table 10.1 – Sample reuse matrix (source: Content Rules)

	User guide	Data sheet	Sales demo slide deck	Webinar	Sales guidance chatbot	Blog article	Support knowledge-base article
Product description (short)	×	×	×	×	×	×	×
Product description (long)			×				
Feature list (collection)		×	×	×		×	
Individual feature (1 item)					×		
Process flow chart	×			×			×
Instructions	×						×
Customer testimonial			×	×		×	

Typically, the best time to build your reuse matrix is while you develop content models for your output types and components. The same analysis process you perform when figuring out standards for those models can inform your decisions about which content to reuse, when, and how.

Some content has specific needs. For example, you might have extremely detailed product specifications, for which content must be reused at the word level. If a widget's maximum height is 5mm, then it's 5mm, no matter where that measurement is used. Your reuse strategy can enable authors to reuse content at the word level for specific, defined types of content, but prohibit them from doing so for other types of content.

There are many possibilities for content reuse. But bear in mind that every new possibility comes with an associated management decision. The greater the variety of reuse types, the more effort you need to plan and manage the strategy. Also, the more reuse you allow, the more important it is for all your content creators to understand and abide by your reuse strategy.

[3] Determine Reuse Management

Content reuse introduces some interesting possibilities into your content management operations.

Storage

Content creators must be able to find reusable content easily. Otherwise, they might assume that no content exists for their given purpose and create redundant content. This new content then requires reviews, approvals, and publishing—all downstream processes that are unnecessary when content is reused.

Your reuse strategy should define where and how you store reusable content. Many content management systems (CMSs) use a folder structure. Others use a category or taxonomy structure. Some use both.

We find that creating reuse libraries at various levels of your repository can make managing reusable content much easier. We recommend creating a common reuse library at the top level of the repository, where everyone can access it. Then, create more specific reuse libraries within product lines. Within each library, you can include folders for different content types.

Here is a simplified example:

1. Repository
   a. Reuse Library for Common Content
      i. Legal
      ii. Glossary
      iii. Hazard Statements
      iv. Common Icons
   b. Product A
      i. Reuse Library for Product A
         1. Feature Descriptions
         2. Icons
      ii. Glossary
      iii. Hazard Statements
      iv. Common Icons
   c. Product B
      i. Reuse Library for Product B
         1. Feature Descriptions
         2. Icons
   d. Product C
      i. Reuse Library for Product C
         1. Feature Descriptions
         2. Icons

Various tools and services are available to identify reuse opportunities and create reuse libraries. These tools are quite valuable. Automating the process of locating reuse opportunities eliminates a lot of time and effort, compared with manually searching for those opportunities. In addition, these tools provide reuse reports that offer insight into how your content creators are writing. You can use this information to standardize content across all five dimensions.

We advise against structuring your repository or reuse libraries by organizational silo. If you organize components and elements based on content rather than department, your CMS can weather any number of business reorganizations.

Any CMS that enables reuse will also track all the relationships across your content. You cannot lose content. Still, keeping it organized is a better option than making content creators hunt throughout a repository every time they need to find something.

Cadence

You can create reusable content outside the cadence of a regular release or marketing campaign. Your reuse strategy should define which types of content can be created outside of such cycles.

For example, you might create a collection of product tips. Each product tip has an image and a short paragraph describing the tip. A tip can help customers get more benefit from a feature or can reveal an advanced capability that most customers don't know about. These product tips can then be reused in product documentation, training, and support. They can also be pulled into eNewsletters, social posts, and web pages.

Other types of content you can create outside of a release include glossary entries, hazard statements, and user-interface label descriptions.

[4] Define Governance

Plan for reuse strategy governance from the beginning. A governance plan can help you avoid the pitfalls that await those who leap into reuse with an entirely ad-hoc approach.

Your reuse governance strategy should define:

- Who can create or revise content that is reused across silos?
- When does a unique component transition to a reuse component? What are the criteria and workflow for such a transition?
- Who owns reusable content: The team that created it, the team that uses it the most, a dedicated reuse manager, or someone else?
- When is reused content retired? Who can retire or archive it?
- What is the process for requesting changes to reused content?
- Who creates and maintains reuse mechanisms or automation?

We recommend answering the *who* questions with a role, rather than an individual or job title. Individuals change, and job titles often do not reflect what that person actually does on any given day. Typical roles include Creator, Owner, Reviewer, Approver, and Project Manager.

Also determine where reused content is stored, who approves it, who maintains it, and any business rules for where the content can and cannot be reused.

Content Development

For content development, provide templates that include reused content. These templates help content creators reuse the right content in the right place. They also reduce or eliminate time spent looking for reusable content (or worse, creating redundant content that must then be reviewed and approved).

A handful of applications are also available to help content creators find reusable content.

Some of these applications work in real-time, while you write. You type a few words, pause, and the reuse engine searches the repository for a match. If it finds something, it shows the suggestion, and you can click to accept it. Other applications are intended for bulk-checking. These applications run on as much content as you can feed them, searching for exact matches (word-for-word duplication) and fuzzy matches (not quite word-for-word, but possibly a close match). The results enable teams to find and resolve small discrepancies in redundant content to create valuable reusable content assets.

Some tools rely on translation memory to locate reused content. Translation memory is a database of all sentences that have been translated. Be careful if you use a translation memory database as your sentence locator. Often, translation memories become riddled with several sentences that mean the same thing yet use slightly different words. When this happens, you need to decide which sentence is accurate. You then need to change all the almost-accurate sentences to match.

Content Delivery

For content delivery, personalization platforms provide various levels of automation for matching content to customer. You'll need to consider both the CMS and content delivery platform to ensure that you provide enough semantic information to support the automation.

The more content reuse you automate, the more scalable your personalization efforts.

Summary

Content reuse enables you to provide a personalized experience while maintaining a consistent message. You cannot handcraft personalized experiences at scale.

A reuse strategy is particularly important when multiple teams (including marketing agencies, outsourced writers, or other external contributors) create your content. Everyone needs to know where to find reusable content, how to create reusable content, and which unique content they'll need to create to meet your business goals.

Your reuse strategy is where you really begin to see the results of your work standardizing content. This is where consistency in your words, sentences, paragraphs, components, and output types pays off.

CHAPTER 11

Taxonomy and Metadata

To provide a personalized content experience, you need to break your content into small, nimble, manageable chunks. You need to write them so that they can be seamlessly pieced together in different ways to create personalized outputs at the point of delivery.

But—and this is important—no matter how wonderfully written and easy to integrate these content chunks are, the entire system is useless if you cannot locate exactly the chunk you need when you need it. This is true regardless of whether people or machines are doing the searching.

Many companies start their personalization journey by developing robust, detailed taxonomies with hundreds of metadata tags. We understand this decision.

Unfortunately, companies often develop their strategy from a content delivery perspective only. To deliver personalized experiences at scale, begin your taxonomy and metadata journey further upstream at the source of the content, where it is created and managed.

What Is a Taxonomy?

A *taxonomy* is a system of organization or classification. Our daily lives contain many examples of taxonomies:

- Biological organisms
- Library systems
- Supermarket layouts

Even at home, most of us naturally create taxonomies to locate our stuff. Consider your kitchen. In all likelihood, you have:

- Plates in the same cabinet
- Silverware in the same drawer (imagine if you had teaspoons in one drawer and soup spoons in another, forks in a third...)
- Knives in a knife block
- Spices in the same cabinet or drawer

Our personal taxonomies might not make any sense to someone else. But as long as we know our individual systems, we can usually locate whatever we need. (Except, of course, when one of the children empties the dishwasher. But that is a problem for a different book.)

Let's look at some of the taxonomies that we interact with.

Biological Taxonomy

In biology, the Linnaean taxonomy classifies all organisms into a *binomial nomenclature*. A binomial nomenclature uses genus and species identifiers to name each organism. For example, you can find your dog Fido under:[1]

```
Animalia
     Chordata
          Mammalia
               Carnivora
                    Canidae
                         Canis
                              Canis lupus
                                   Canis lupus familiaris
```

Canis Lupus familiaris is the binomial nomenclature for *Homo sapiens'* best friend (the dog).

Library Taxonomies

All libraries use a taxonomy to organize books and other materials. Three taxonomies commonly used by libraries are:

- **Dewey Decimal:** used in public libraries
- **Library of Congress:** used in academic libraries
- **Superintendent of Documents:** used in academic libraries with federal depositories

You can have the same books available in each system. However, the way the books are organized is different depending on which system you use.

---

[1] "Canis lupus familiaris" (Wilson 2005)

Dewey Decimal

In the Dewey Decimal system, books are grouped into 10 subject areas. Each group is assigned 100 numbers.[2]

Number	Subject Area
000–099	General Works
100–199	Philosophy and Psychology
200–299	Religion
300–399	Social Sciences
400–499	Language
500–599	Natural Sciences and Mathematics
600–699	Technology
700–799	The Arts
800–899	Literature and Rhetoric
900–999	History, Biography, and Geography

Library of Congress

The Library of Congress system divides books into categories, lettered A through Z.[3]

Letter	Subject
A	General Works
B	Philosophy, Psychology, Religion
C	Auxiliary Sciences of History

---

[2] "Introduction to the Dewey Decimal Classification" (OCLC Online Computer Library Center 2011)

[3] *Library of Congress Classification Outline* (Library of Congress)

Letter	Subject
D	World History and History of Europe, Asia, Africa, Australia, New Zealand, etc.
E	History of the Americas
F	History of the Americas
G	Geography, Anthropology, Recreation
H	Social Sciences
J	Political Science
K	Law
L	Education
M	Music and Books on Music
N	Fine Arts
P	Language and Literature
Q	Science
R	Medicine
S	Agriculture
T	Technology
U	Military Science
V	Naval Science
Z	Bibliography, Library Science, and General Information Resources

Each subject area is further defined into several subclasses.

Superintendent of Documents

The Superintendent of Documents system organizes materials by governmental publishing agency rather than by subject. The following table includes some examples.[4]

Letter	Agency
A	Agriculture
AE	National Archives and Records Administration
B	Broadcasting Board of Governors
C	Commerce Department
CC	Federal Communications Commission
CR	Civil Rights Commission
D	Defense Department
E	Energy Department
ED	Education Department
EP	Environmental Protection Agency
FA	Fine Arts Commission
FCA	Farm Credit Administration
FHF	Federal Housing Financing Board
FM	Federal Mediation and Conciliation Service
FMC	Federal Maritime Commission
FR	Federal Reserve System Board of Governors
FT	Federal Trade Commission
FTZ	Foreign-Trade Zones Board

---

[4] "Superintendent of Documents (SuDocs) Classification Scheme" (Federal Depository Library Program 2010)

After the alphabetical symbol, the Superintendent of Documents system further classifies information using integers and punctuation marks.

Any book can be housed in multiple libraries, each using a different taxonomy to organize that book. That same book can be found using a different set of metadata (call numbers), depending on which library taxonomy is in use.

Using Library Taxonomies

Suppose you want to learn more about Fido.

If you want to find books about Fido in a public library that uses the Dewey Decimal system, you find Fido under:[5]

600	Technology
630	Agriculture and related technologies
636	Animal husbandry
636.7	**Dogs**
636.8	Cats

And if you use the Library of Congress classification, you find Fido under:[6]

- S   – AGRICULTURE
  - SF – Animal Culture
    - SF411-459 – Pets
      - **SF421-440.2 Dogs.** Dog racing
      - SF441-450 Cats
      - SF451-455 Rabbits and hares
      - SF456-458.83 Fishes. Aquariums
      - SF459-459 Other animals

Of course, books about dogs aren't the only things that can be organized in multiple ways.

---

[5] "Introduction to the Dewey Decimal Classification" (OCLC Online Computer Library Center 2011)

[6] *Library of Congress Classification Outline* (Library of Congress)

Supermarket Taxonomies

Imagine if supermarkets didn't use a physical taxonomy to govern the placement of items. If you wanted to buy a can of baked beans, how would you know where to look for it? It might be next to the bread, or perhaps next to the orange juice. But because supermarket layouts use a taxonomy, you know you can find baked beans with other canned foods.

Yet have you ever noticed that sometimes the same food item is displayed in more than one place in a supermarket? Take those baked beans. Not only are they in the canned foods section, but the same type of baked beans might also be in the ethnic foods section. Having the same type of beans in both places makes it even easier to find the beans that you want.

Supermarket taxonomies don't end with the grouping (or groupings) of similar types of food. Most U.S. supermarkets also use a similar physical taxonomy.

This physical taxonomy groups:

- Produce on one side
- Deli counter and bakery products on the other side
- Meat, poultry, and fish along the back wall
- Canned goods near the center aisles
- Frozen food in the center aisles
- Candy, magazines, and impulse buys at the checkout stand

Individual shelves at the supermarket are also organized in a taxonomy:

- Best-selling brands in the center set of shelves
- Bulk and store brands on the bottom shelves
- Gourmet and regional brands on the top shelves

And of course, dairy products are organized according to expiration date, with the oldest products in front and newer products in back.

We are so accustomed to these typical taxonomies that any supermarket with a different layout (Trader Joe's, I'm looking at you) can be very disorienting. Items in those stores can be difficult to locate. After all, who puts the cookies above the frozen fish?

Taxonomy and Enterprise Content

Just as a taxonomy is used to organize books, animals, and food, a robust and carefully designed taxonomy is important to organize content. An enterprise taxonomy is needed to:

- Organize a large volume of information
- Improve search speed and accuracy
- Enforce consistency
- Increase content usability
- Enable content reuse

If you cannot find your content, you cannot reuse your content. If your systems cannot find your content, you cannot automate content reuse. Automating reuse is a key component of delivering personalized experiences at scale.

Your content delivery system matches the information it has about your customer with the information it has about your content. It makes this match so quickly that your customer doesn't know the search-and-retrieval process is happening in the background.

You cannot rely on full-text searching to deliver the right content to the right person at the right time. Full-text search takes too long. Even Google Search doesn't do a full-text search of the entire internet every time you search for something. Google software continually indexes and tags content behind the scenes. When you enter a search phrase such as "dog park near Regina's house," Google searches just the content that it has already identified as "dog park" and "near

Regina's house." It doesn't have to search any other content because it has categorized the content ahead of time.

The more consistently you classify and tag your content where it is created and managed, the more accurate your content is delivered after it is published.

In most companies, content creators assign taxonomy categories when they create the content. In some companies, all content goes through a taxonomist for review before it is published, the same way written content is be reviewed by an editor.

To successfully create personalized experiences at scale, you need a well-thought-out taxonomy. You need a system that everyone uses consistently (without fail) to organize the content. The more content and content creators you have, the more critical it is to have a planned taxonomy.

What Is Metadata?

Technically speaking, metadata is data about data. Metadata is an individual marker, also called a *tag*, that is assigned to a piece of information to help describe and locate that data. Metadata is the tool that you use to organize your taxonomy.

For example, consider clothing.

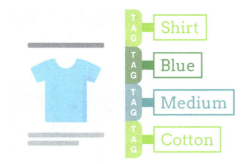

This illustration shows clothing that is tagged with identifying metadata. By looking at the tags, you know that the article of clothing is a blue shirt, size medium, made of cotton. Because the shirt is tagged with this metadata, you can search for any category—type, color, size, or material— and find this shirt in the search results. If you use all four tags in your search, you can find this shirt even more quickly.

You must carefully plan and execute metadata, just as you do your taxonomy. Also, design metadata to be scalable, to account for new products, services, and other things you do not yet know about.

Types of Metadata

There are three main types of metadata:

- **Structural:** describes the relationship of the asset to another asset (e.g., chapters, indexes, sections)
- **Administrative:** describes information about the source of the asset (e.g., file permissions, creation date, revision number)
- **Descriptive:** describes the asset itself (e.g., subject matter, product line, intended audience)

Metadata and Tagging

Applying metadata is also known as *tagging*. When you apply metadata to content, each piece of metadata becomes a tag that can be used to locate that content during a search. Searches are performed by humans or machines. In essence, a personalized experience is a series of search-and-retrieve operations performed by machines at faster-than-human speeds.

Metadata is used to find content both internally and externally:

- **Internal metadata:** provides tags that you can use to search your internal repository for existing content. For example, you can search for content that has not yet been published and that was created by a specific employee within a defined date range.
- **External metadata:** is used by customers to find content.

In essence, metadata tags do the same thing—identify content—whether used internally or externally. The only difference is that internal metadata is used by people who create source content, and external metadata is used by people who consume published content.

Companies often focus more attention on external metadata (for external search) than on internal metadata. This is a mistake. Without internal metadata, it is difficult to reuse content effectively and at scale. After all, to reuse the content, you first need to find the content.

If content creators cannot find a content component quickly, they will likely write a new one. And that completely defeats the purpose of content reuse at scale. Remember, humans have very short attention spans: approximately 10 seconds. So, we have 10 seconds to make sure that content creators can find the content they need before they give up and write it all over again.

Internal and external tags are often identical. However, we sometimes use different search terms internally. For example, many companies assign placeholder names for products that are in development. When the product nears release, the official product name becomes standard.

> Val once worked for a company that developed two versions of a server. Internally, one was named *Barney* and the other *Fred*. When the product was released, those names were replaced with official, branded names (we'll call them *System 500* and *System 1000*).
>
> While Val's writers were creating content for Barney and Fred, they used the internal tags Barney and Fred. After the products were released, that content used the external tags System 500 and System 1000.
>
> The internal tags were not exposed outside the company. The external tags were maintained both internally and externally, for convenience.

Bringing It All Together

Let's return to our housing analogy to illustrate how our taxonomy helps us classify content for a personalized experience.

When we personalize a home in a planned housing community, we present buyers with a taxonomy that organizes their choices into several categories.

Here are just a few of those categories:

Rooms | Style | Finishes | Hardware | Flooring |

Each category has associated options, described by metadata tags:

Room	Style	Finish	Hardware	Flooring
Attic	CapeCod	Brass	Deadbolt	Carpet
Basement	Colonial	Bronze	Door	Cork
Bathroom	Cottage	Nickel	Doorknob	Laminate
Bedroom	CountryFrench	Chrome	Fastener	Tile
Bonus	Craftsman	Crystal	Faucet	Wood
Dining	Mediterranean	Paint	Hinge	
Family	Modern	Stain	Hook	
Garage	Ranch	Wallpaper	Lock	
Kitchen	Tudor			
Living Room	Victorian			
Study				
Utility				

Our taxonomy allows us to tag our content from more than one category. For example, here is an image that is categorized in the following categories:

- Hardware = Doorknob
- Style = Victorian
- Finish = Crystal

Here is an image that is categorized in the following categories:

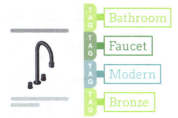

- Room = Bathroom
- Hardware = Faucet
- Style = Modern
- Finish = Bronze

We might also have an image that fits more than one way into a single category. For example, we might want that crystal doorknob to be used in any of the more private rooms in the house. We could categorize it like this:

- Room = Bathroom, Bedroom, Study
- Hardware = Doorknob
- Style = Victorian
- Finish = Crystal

By tagging our source content in this way, our content creators can easily find exactly the right image to include in their content. Those tags can travel with the content all the way through the publishing process. Once published, those tags become available to the content delivery platform. At that point, a customer who previously explored our Victorian styles can be shown our Victorian crystal doorknob at the top of our doorknob web page.

Another customer visiting us at the same time—even from the same room—might be shown a Cape Code doorknob in that top slot, based on their own previous interactions with doorknobs or home styles.

The Relationship Between Standardization, Taxonomy, and Metadata

In a standardized environment where content reuse is a primary part of the strategy, what is the first thing a content creator needs to do? They must determine whether content already exists. If they cannot quickly (within one or two searches) find an existing piece of content, they will simply create a new chunk—potentially one that is almost identical to an existing chunk.

This is what we call a *standardized authoring fail*. It happens when you have multiple, nearly identical chunks of content. The failure is multifaceted:

- Writers waste time creating the same content over and over again.
- Systems don't know which piece of content to publish for a given person's experience.
- Almost identical chunks, written at different times, contain different versions of the same information.

It is simply a mess.

Similarly, when a customer comes to your website and searches for information, you get only one or two chances to serve up the exact information the customer is looking for. If you don't provide relevant information within that time, the customer will move on to your competition.

Internal and external search are critical components for delivering personalized experiences. Done right, search is like a well-choreographed ballet. Do it wrong, and you'll have customers and potential customers leaving your site in droves.

The first key to preventing a standardized authoring fail is a well-defined taxonomy. The second key is consistent tagging of content, using metadata that makes sense.

I am convinced beyond a shadow of a doubt that the first thing people need to do is buy into predictable, machine-processable content encodings with rich, self-describing semantics. If a machine cannot programmatically differentiate content containers and their relationships—if there's not enough semantic information aside from the actual text itself—it will result in utter failure at scale.

—Mike Iantosca, Enterprise Content Strategist

How Big Can You Scale?

You can see from our (very) simple example just how powerful your taxonomy and metadata standards can be for both organizing content internally and setting it up for personalized delivery.

Regina once helped a company in the travel industry develop a massive enterprise taxonomy that accounted for every possible variation of someone's travel plans. This company did not create content itself. Instead, it ingested massive amounts of content created by other companies – content created with no thoughts of structure or standards. Our customer's proprietary software analyzed the content, tagged it, and then delivered the tagged content to its subscribers.

At the time, "alternative travel" was looming as a viable competitor to established hotels. Our customer wanted to update their taxonomy and metadata standards to support content far beyond the old-school hotel tags such as *pool* and *gym*. They wanted to encompass everything from AirBnB to glamping to Survivor-esque adventure tours in the world's most remote locations.

This customer did not need to consider the capacity of human content creators in developing this taxonomy. Their software did not get overwhelmed by hundreds of categories and thousands upon thousands of metadata tags. The end result? Their subscribers are able to provide highly personalized experiences to customers who have no idea what it takes to enable them to do something as simple as book a hotel room.

Developing Your Taxonomy

Here are some basic principles to keep in mind when developing your enterprise taxonomy.

- **Semantics:** Categories must have meaningful names. Category names such as *Category 1*, *Category 2*, and *Category 3* help nobody. Category names such as *Product*, *Audience*, and *VendorID* are much more useful. Real words also help humans understand the tags at a glance, even if machines just treat each name as a string.
- **Uniqueness:** For most systems, each taxonomy category and metadata tag must have a unique name. Most companies take advantage of capital letters or underscores to help humans read the tags. For example, we can understand that FrenchCountry and French_Country both mean French Country. You'll need to verify your system capabilities and develop a naming standard that works throughout the content ecosystem.
- **Hierarchy:** Traditional taxonomies developed to organize physical items often have a defined hierarchy. For digital content, you may not need as much hierarchy as more traditional taxonomies suggest.

- **Usage:** Each taxonomy category contains a set of metadata tags. Will you apply just one tag from that category? Or can you apply many? For example, our personalized home taxonomy allows the image of our doorknob to have multiple tags from the Room category.
- **Scope:** Ultimately, your enterprise taxonomy needs to be just that: a taxonomy that supports the entire enterprise. Most companies have a set of categories that apply to content created throughout the company plus additional sets that apply only to one batch of content. For example, learning content often has a need to specify the intended audience of a component (e.g., facilitator, participant, or novice) in a way that customer support content does not.

Knowledge Graphs

The next level of organization is a *knowledge graph.*

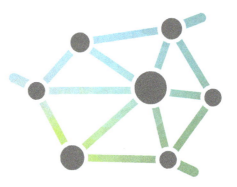

A knowledge graph shows relationships among entities. Companies that have vast amounts of content have started using knowledge graphs to represent their content and the connections within it.

Knowledge graphs use *ontologies.* Think of an ontology as a taxonomy of taxonomies: a group of taxonomies that are tied together by overlapping metadata. A knowledge graph relates the information in different content taxonomies.

For example, suppose your technical documentation group has a taxonomy that contains all the information it needs to produce a particular output. Your marketing group has its own taxonomy that describes the content it produces. But both groups consistently tag content, using the same metadata system within the group.

Some of the content that each group creates is also used by the other group. A knowledge graph ontology links the two taxonomies and shows how the content in each works together.

With content teams often working in different systems, it hasn't been historically possible to combine taxonomies into a single enterprise solution, much less draw on a single enterprise knowledge graph. However, many component content management systems now integrate with applications that are designed to manage taxonomies and to apply metadata to content. Content teams can thus share a taxonomy and build knowledge graphs even before they can share content.

Automation is Coming Soon

To deliver personalized experiences at scale, the delivery mechanism needs to be automated. That means machines, not people, need to locate the content that customers need at the moment they need it.

For now, systems rely on robust, standardized, and consistently applied metadata to automate this process. To support this automation, every content creator in your enterprise must use the same metadata for the same things. And that metadata needs to be applied without fail. Otherwise, automated systems cannot find the right content, and you cannot successfully personalize the customer experience.

We think the need for humans to develop a complete metadata model will start to diminish in the future, as AI systems become more commonplace. AI systems will develop their own ever-evolving metadata models to increase speed and accuracy of search-and-retrieval. AI will also perform full-text searches, even handling synonyms and misspellings, at blazingly fast speeds.

You'll be able to use your enterprise taxonomy and metadata model to train the machine. But for now, taxonomy and metadata hold the key to findability and the delivery of personalized experiences at scale.

Summary

Taxonomy and metadata enable reusability and automation, both of which are critical to delivering personalized experiences. By providing mechanisms to organize content and tag it for retrieval, both systems enable content creators—and customers—to find content more easily. Taxonomy and metadata need to be standardized across your company and applied consistently for each and every chunk of content.

CHAPTER 12
Standardizing Media

Just like any other content, media must be created to standards. Standardization greatly increases the opportunities for reuse. Considering how much media creation costs in time, talent, and resources, it would be a shame for media to be used only once.

What Is Media?

Media is a general term for any non-textual content components, including:

- Images
- Charts and graphs
- Infographics
- Videos
- Audio tracks
- Animations

We're willing to bet 1,000 pretend dollars that your company has a whole library of media that you can't reuse. Maybe the media was created by an outside agency, used for a few months in a marketing campaign, and then archived (or worse, deleted), never to be seen again. Maybe an image contains text that makes the image unsuitable for a different experience. Maybe you have a 10-minute eLearning video gathering virtual dust.

The barriers to reusing media often relate more to organizational silos than to the content itself. In many of the companies we've worked with, product documentation and customer support cannot access media created by marketing or training. But everything you do to standardize, organize, and tag written content applies to media assets as well.

Creating media standards that optimize media for personalization and reuse can go a long way toward ensuring that you get the most value from every media asset. Standards also help guide agencies and outsourced developers to create media that meets your needs without as many reviews and iterations.

More Media!

It is long past time that your company's content creators focus more on media and less on text. (We say this knowing full well that, as writers, we too are most comfortable with text.) After all, if customers ask for a series of short how-to videos and you hand them a 300-page PDF, you aren't optimizing the customer experience.

Our clients often tell us that their customers have been asking for more media and less text, particularly in product documentation, for years. And yet, although technical content creators often acknowledge that the customers' needs are not being met, most technical publications are still delivered as PDFs.

Meanwhile, marketing organizations are creating gorgeous product photos and illustrations. Training departments are creating animated videos with voiceovers and transcripts.

Reasons for technical content creators' attachment to PDF include:

- **Budget:** PDFs are cheap to produce.
- **Tradition:** The old guard is comfortable and doesn't want to learn new things.
- **Capacity:** Nobody has time to develop a new process.

The good news? Standardizing content can help content creators recapture time, which they can then use to develop the media that customers want.

What to Standardize

Let us reassure you: We aren't about to suggest that you take all the creativity out of your media. Far from it. But the visual and aural content that your company produces needs to look and sound like it comes from one voice.

To return to one of our previous examples, if your company voice is like the State Farm friendly neighbor, you can't expect Geico-quirky videos to blend into a cohesive personalized experience.

To optimize your delivery of personalized experiences at scale, you need to create media standards in two areas: content and asset management.

Media Content Standards

Most of our clients already have some content standards in place to guide the look and feel of customer-facing media. Content standards define what is and what is not allowed in the media itself. For example:

- **Brand standards** describe color palettes, fonts, logo usage, taglines, and typography.
- **Video standards** describe resolutions, aspect ratios, intros and outros, and text overlays.
- **Iconography standards** describe the visual language of symbols and icons.

Content standards help to ensure that visual elements fit together when combined into an output or personalized experience. These standards build your brand and help content creators across different teams develop media content that looks like it came from one place. They are your media style guide.

Unfortunately, we rarely find that *all* content creators in a company both know about and have access to the company's media standards. Without knowledge of and access to your standards, teams end up creating illustrations and videos that do not conform to a single visual language.

Media Asset Management Standards

We often work with teams that start out with an entirely manual process of managing their media assets. This process usually involves network folders, file naming conventions, and everyone's best attempts to comply with procedures and standards.

And more companies than you might think have no media management at all. Images, charts, and videos are handled on an entirely ad-hoc basis. Each individual creates media as needed and produces the assets with whatever format, size, resolution, and other attributes seem best to them at the time. There is little or no attempt to maintain raw source files or reuse the asset in other content. In some cases, image and video source files are even stored on individual laptops rather than a network drive.

One internal challenge to media reuse is that you can't simply find existing media via a full-text search. With textual content, you can search for important words to find what you need—but media is non-textual. With videos, you can work around this limitation by using transcription software to convert audio to text, and then store transcripts with videos. But with images, you can't even do that. And if your own content creators can't find the right media content, how will customers (or your automation platform)?

Fortunately, you can use taxonomy and metadata to help content creators find exactly what they need, standardize your media content, *and* deliver personalized experiences. Organize your media assets according to your company taxonomy. Use the same metadata on media as you use on other components.

You can extend your taxonomy with media-specific categories, such as resolution or aspect ratio. You can also take advantage of free-text fields for descriptions and other searchable text.

In recent years, more and more of our clients have considered a Digital Asset Management (DAM) system—a system for storing and tagging images for ease of identification and retrieval—as they develop their company taxonomy. However, unless all content creators follow a standard method for tagging images, including any and all required metadata, DAMs can create a good deal of frustration.

In an ideal world we would gather all our media content into one centralized asset manager that all our content teams could access. We would create media according to standards and carefully curate it so that the right media was always used in the right output types at the right time for the right experience.

So, how do we move from a manual system to our ideal world? The first step is standardization. Complete, consistent commitment to standards across every team, every type of content, and every asset-management system is critical to including media in your delivery of personalized experiences at scale. Ensuring that your taxonomy includes the categories and metadata necessary to help content creators leverage media assets goes a long way toward reusing content across your company.

A True Story

Back when Val was manager of technical training for a networking company, she oversaw the creation and management of illustrations for both training content and documentation. The illustrations were primarily line drawings of network hardware. They were very detailed, showing ports and pins and all sorts of angles. Some even showed hands holding a printed circuit board or replacing a fan subsystem. It was very important that each diagram be re-usable for training, documentation, lab exercises, and so on.

At that time (the Dark Ages), the only available content management technique was shared network folders. Because they had no metadata, tagging, or even the concept of a content management system, Val and her team used the only approach they could think of: a unique name_number moniker.

Every illustration had a unique filename that contained information about the image. For example, the 132nd image created about **Widget AB** used filename AB_132. For ease of use, the team also placed the filename in 4-point font at the bottom of each diagram.

All the images were stored in a series of folders, one per widget type. The theory was that if you needed an image of **Widget AB**, you'd look in the AB folder. If you needed an image of **Widget CZ**, you'd look in the CZ folder. The numbers were simply assigned sequentially. If an image needed to be modified because a port on **Widget AB** changed or was moved to a different location, a revision letter was added to the filename; for example, AB_132a.

As you might imagine, this system worked well at first. When the company had only two products, keeping track of the images was simple. However, as the company grew, it created more models and units. The illustrators created more and more drawings. And the nomenclature for the numbering system exploded. Reuse by filename simply did not scale.

That story may be from "long ago," but not all companies have let go of the filename-and-network-drive method of managing media. As recently as 2020, Content Rules has helped customers develop media management strategies in exactly that type of environment. In developing the filenames and folder structures, we helped them determine their most important information for finding media content in the future when they do—eventually—get their DAM requests approved.

How Media (Mis)management Affects Personalization

In many companies, each content silo creates, stores, and manages its own images. Marketing images are rarely used in the knowledge base, technical documentation doesn't usually share images with training, and so on. Not only does this silo-ed way of doing things create a lot of duplicated effort, your company ends up with images that vary greatly in quality and style.

Back when customers only saw images in the context of full documents or on your website, this inconsistency didn't matter. But as you move from delivering *documents* and *static web pages* to delivering *personalized experiences*, customers will see—and notice—inconsistent images at different points in their journey. At best, they'll recognize the media as being part of your company's brand. At worst, they'll be turned away by the low quality of the experience or be confused by the inconsistency.

How many times have you encountered professionally created, engaging media in product marketing content, only to be confronted by amateurish block diagrams in the product documentation? This is exactly what we mean when we talk about mismanaging images and media.

How to Standardize Media

These best practices can help you create media that helps your personalization efforts.

1. Separate text from media.
2. Develop content models for media.
3. Use your company taxonomy to organize media.
4. Tag media with metadata.

Let's discuss each step in more detail.

[1] Separate Text from Media

Text in images and video is problematic. Embedded text cannot be changed on the fly. It cannot be easily localized, much less personalized. Embedding text in an image or video often limits the media to a particular region, language, season, product, campaign, or time and place.

Instead, create images and videos with the assumption that any necessary text will be provided separately.

Think of an image that needs callouts. Instead of putting the text directly into the image, label the relevant parts of the image with numbers such as 1, 2, 3. Then provide the text descriptions separately.

Here's an example of an image with labels followed by text descriptions. The same image can be used for several purposes, if you manage the text separately from the image.

1. Ear
2. Snout

The same approach works for video. Instead of hard coding text into a video asset, use software to generate transcripts and provide captions automatically. You can find applications that transcribe the audio and create links to specific parts of the video. You can then localize the transcripts. Captioning is automatically displayed in the customer's preferred language.

[2] Develop Content Models for Media

If you thought content models were only for writing, think again. A successful company content strategy includes models for media, for the same reasons you create models for output types and components. Here's how to create models for media:

1. Identify the types of media in your content (e.g., decorative images, tutorial videos).
2. Define the purpose of each media type (e.g., demo product, engage customer).
3. Develop models for similar media (e.g., all how-to videos include no more than 10 seconds of context before reaching step 1).
4. Develop guidelines for media usage (e.g., no more than one video per product web page).
5. Leverage your taxonomy and metadata strategy to make images easier to manage.
6. Design an image workflow that automates media-related tasks that can be automated and streamlines media-related tasks that can't.
7. Provide ongoing governance and grooming.

Every media asset has a job to do. To do that job, the media asset must contain certain elements.

For example, the content model for a screenshot defines criteria for when to include labels and which type of labels to use (e.g., circles or squares, numbers or letters). The content model for a tutorial video defines which reusable clip to use for the intro, the maximum length of a clip covering any one feature, and which reusable clip to use for the outro.

At a minimum, each media asset needs a title and a description for internal use. Even if your customers do not see the title of an image, your content creators need the titles to help them search and find the right image for their content.

[3] Use Your Taxonomy to Organize Media

The same taxonomy categories you developed for your text content apply to your media content. From a taxonomy standpoint, there is no difference between a text component and a media component. Extend your taxonomy to include additional categories and metadata for media that are not relevant for text.

If you use a DAM to manage media, you probably have robust taxonomy capabilities for organizing files. Wherever possible, take advantage of controlled vocabularies to provide consistent tagging for your media. If you manage media in a file system, you can use your enterprise taxonomy as a basis for your folder structure.

Because full-text search is not an option with images, ensure that content teams organize media so that content creators can easily find and use it. If you extract text transcripts from videos, store the transcript with the video to help content creators find the video more easily.

[4] Tag Media with Metadata

Tag media with metadata so that humans and systems can find the right media asset. Humans need metadata to help with search filters and results. Systems need metadata to match the media with the experience.

Arguably, tagging your media in a consistent way is even more important than tagging your written content. With written content, you usually have the option of full-text search. That means content creators can usually find the content they need (eventually). After the content is published, your customers also have a hope of finding it (even though you cannot provide personalized experiences at scale without consistent tags for all your content).

We recommend tagging media the way you tag any other component. Start with the same taxonomy categories and metadata tags that you developed for your content management system. Then, add any categories that help identify the media, for example:

- Format (.png, .jpg, .mp4)
- Media type (illustration, screenshot, video)
- Resolution (low, high)

You might also need categories to enforce usage rules around media, such as:

- Regional restriction (used only in defined markets or countries)
- Demographic (age, gender, profession)
- Licensing (accreditation requirements, expiration dates)

Summary

We want all our content to work well together, regardless of who creates, assembles, publishes, or receives it. We want product content to benefit from the professionalism of the images that marketing creates, and we want support, training, and documentation to provide a harmonious customer experience.

We don't want our customers to find an image in product documentation that contradicts what they encountered in training or that differs from what they saw in our marketing materials.

When you define standards and develop models for your media content, your media assets become reusable across business silos and content delivery platforms. Remember, access to all your content, regardless of format, is how you create personalized experiences at scale.

CHAPTER 13

System Requirements

The tool is never first. A company needs to have done their analysis and design and know what the requirements are before they can even identify that they need new tools or that they need to modify existing tools.
—Ann Rockley, CEO, The Rockley Group

This book is not about systems. But it's hard to discuss the management of standardized content and delivery of personalized experiences without thinking about the tools and systems you need.

In our experience, focusing on systems first is one of the primary reason companies fail in their attempt to deliver personalized experiences at scale.

This focus on software seems to work in two ways. Someone in the organization thinks:

- "We have software X and it can do Y. Let's do it!"
- "We want to do Y. Let's buy software X and do it!"

In either case, they can sometimes make software X work long enough to complete a pilot project. But when they attempt to scale that project across the company (or even invite just one additional organization to participate), the number of workarounds, manual efforts, and "We'll fix it later" events prevents them from succeeding.

Part of the problem is that such organizations are thinking about the wrong systems. They're starting from the end, with delivered content. Starting from the end forces your organization to make decisions based on how you will publish content. And regardless of how you publish, you must start at the beginning: by focusing on your source content.

We cannot say this enough times. You must standardize your content first, and manage it in a standardized content ecosystem. You cannot simply apply personalization to the end product and expect positive results.

The Elements of the Content Ecosystem

Here's a quick look at the types of software you need to create and manage standardized content.

To deliver personalized experiences at scale, you need *at least* the following elements:

- Component content management system (CCMS)
- Digital asset manager (DAM)
- Personalized content delivery platform

Component Content Management System

A CCMS provides a single source of truth for standardized content. The difference between a content management system (CMS) and a CCMS is in the level of content granularity.

Most CMSs manage content at the document level. A CCMS manages content at a much finer level of granularity: the component level. With a CCMS, you can mix and match small pieces of content in any order you need to create a personalized experience.

Figure 13.1 – CCMS ecosystem

Figure 13.1 shows an ecosystem that uses a CCMS containing the following basic elements:

1. **Components:** Content creators write content as nimble, focused components. Typically, they create this content in one of several authoring tools that connect to a repository. (Some CCMSs include a built-in authoring tool.)

2. **Metadata:** Components are tagged with metadata to categorize their content and to support search and automation.

3. **CCMS with taxonomy:** Content is stored in a repository and organized using a company-standard taxonomy and metadata tags. The repository stores components and assemblies of components. The repository also maintains the relationships between components and tracks component versions, revisions, and languages.

4. **Localization:** The repository stores localized components alongside source components. This integration enables you to write content once, translate it once, and then reuse and publish it as many times and in as many formats as you need.

5. **Content optimization:** Terminology, grammar, style, tone, and other content quality checks happen as part of the content development process.

6. **Publishing engine:** Content creators and automated systems assemble content components for various output types. The publishing engine applies the appropriate stylesheets to add layout and design for each output and channel. You can publish content in the source language, or you can publish translated content. You can also design any number of output formats to accommodate text expansion, bidirectional languages, and so on.

7. **Channels:** Content is published for delivery in any channel. Metadata tags support dynamic, intelligent delivery to provide personalized experiences throughout the customer journey. Some content can be published directly to a content delivery platform, such as a dynamic delivery engine, website, email manager, or mobile app. Other content might need to be published to a staging area before being delivered.

These are just the basics. Your content management scenario becomes even more robust when you integrate your CCMS with other systems.

For example, one of our clients in the high-tech space wanted to integrate its product information management system and content management platform. The product information management system was the company's system of record for things like the length, width, and depth of each hardware product. This system managed all the details that are crucial to manufacture any product. One typo—one "1 mm" that should have been "1 μm"—could doom an entire product run.

Instead of having content creators replicate information in their specifications and product documentation, the company began to design and develop integrations so that detailed data flowed automatically into the content from a single source of truth. The primary motivation was accuracy in production; production errors could cost millions of dollars.[1] That detailed data would also flow downstream to support personalized experiences in documentation, support, and technical marketing.

A basic CCMS can be one system or several integrated systems. Some companies connect their CCMS to a taxonomy management system (TMS) to centrally store and share a taxonomy across all connected systems. A connector in the CCMS reads values from the TMS, providing a term list for CCMS customers. When the TMS is updated, so is the CCMS. Terms can also be filtered based on customer role, department, or other criteria, to avoid an overwhelming number of available terms. A TMS is optional, but we are seeing more and more companies turn to a centralized solution to manage taxonomy.

Document control systems such as SharePoint, Documentum, and G Drive are *not* CMSs. And they are definitely not CCMSs.

Digital Asset Manager

An enterprise-class DAM system provides a repository for media such as images, video, animation, and audio clips. Think of a DAM as a CCMS that is optimized for large file sizes and content assets for which full-text search is not an option.

Media is one of the most reusable content types and one of the most powerfully personalizable content assets. A good DAM provides robust taxonomy and metadata capabilities that enable content creators to find media quickly and accurately. For automation, you need strict adherence to your taxonomy and metadata standards so that the systems can find the right media to assemble and deliver the right content.

Personalized Content Delivery Platform

A personalized content delivery platform provides dynamic, contextual content delivery at the point that the customer needs it. The delivery platform provides an additional layer of metadata to the content that the CCMS publishes. This metadata helps systems match content to customer.

---

[1] Remember the time NASA lost a Mars probe because it used the English system of measurement early on and forgot to convert the measurements to metric? That's $125 million and a spaceship we'll never see again.

A wide range of personalized content delivery platforms are available. Micro-businesses often use small WordPress plugins that deliver personalized offers to customers based on their behavior on a website. Global behemoths often invest in enterprise-scale personalization engines with full machine learning (ML) capabilities, ever-more-sophisticated artificial intelligence (AI), and access to years of historical customer data plus real-time customer behavior.

Another type of content delivery platform that has been gaining popularity is a *headless CMS*. Using a standard delivery platform, this system often provides templates that predetermine the way content is presented.

A headless CMS is a content delivery platform that uses an API call (often coupled with an open-source query language called GraphQL) to retrieve content and display it however you choose. The benefit of a headless CMS is that it does not lock your content into a particular format or output type when presenting it to a customer. Rather than using an existing architecture, a headless CMS uses code that you write to deliver content in the format you need. Because of this flexibility, headless CMS technology is useful for delivering personalized experience at scale.

Sneaking Standards into the System

Migrating content to a new system is a big endeavor. Make sure you develop a solid business strategy for content before making such an effort. **The system should support the content strategy, not drive it.** However …

Sometimes getting funding to buy a new tool that provides new capabilities is easier than getting funding to develop content standards. A new enterprise solution can provide a pathway for change. As everyone thinks about moving to a new tool, you can "sneak in" changes to content management, content standardization, taxonomy, and workflow. Your reason? "That's just the way we need to do things in the new system."

Executives and stakeholders also are often more receptive to phased deployment strategies. This is especially true when a solid master plan is developed and updated with findings during each phase.

If you absolutely cannot get support for content analysis, standardization, and strategy initiatives, you can still go through the process as part of preparing for a new or upgraded system. Just know that if you want that system to provide the business benefits you need, you cannot shortcut the analysis and strategy process.

You can also insist on incorporating analysis and standardization into the selection of a new system, long before purchase. After all, the investment is a necessary one. Your content standards and decisions determine how you will implement and configure the system.

In a digital content world in which our tools continually become more sophisticated, we often think that tools alone can solve or do everything we need. Whether you think about your system first and consider your content later or think about the content first and consider the system later (our preferred order)—sooner or later, you need to think about the entire ecosystem from start to finish if you want to deliver personalized experiences at scale.

What About XML?

In this book, we avoid sinking into the deep technical details involved in mapping your content models to XML. However, we want to acknowledge that your content architecture should define technical requirements for structuring, tagging, and automating content creation and delivery.

These technical requirements depend on the CMS that you use and on the different types of output formats you publish to.

By structuring your content and using semantic XML to support that structure, you can go a long way toward readying your content for personalization. Delivery systems can even retrieve content based on semantic XML elements and their attributes, without an additional layer of metadata.

You can also use a fairly new technology called Semantic AI to create detailed knowledge graphs, using AI and ML. Knowledge graphs organize and classify large quantities of content, creating an interlinked system that provides better targeted and in-depth search results.

When we work with clients to create content models, we document the technical details in the model. These details include semantic XML elements and attributes if the client is mapping the content to an XML structure.

If client does not use XML, we might include notes about forms, field names, heading styles, or other structural indicators that are available in the client's toolset.

What About Reuse?

You can implement content reuse in several ways. After you determine *which* content will be reused and *where*, you can make decisions about *how*. Your reuse strategy informs which mechanisms you use to support each type of reuse.

Reuse mechanisms depend almost entirely on the capabilities of your CMS and your content delivery platform. Here are some examples of reuse mechanisms that our clients have used:

- Content references
- Variables
- Conditions
- Content mappings
- Metadata tags
- Object properties
- XML attributes
- Integrations

If you can't even make your eyes read that list, you aren't alone. This aspect of reuse involves a lot of collaboration between content creators and content engineers. Content creators determine which types of reuse best suit the business goals. Then the content engineers figure out which mechanisms can support those goals.

If your current toolset does not provide the mechanisms that best suit your strategy, you can decide whether to use workarounds, wait for an updated system, or reuse content in some other way.

How you configure the system and which mechanisms you use depends heavily on your system capabilities, how much influence your team has over systems administration or IT, and the amount of content to be reused in this way.

Reusing content can happen automatically during content development, during content delivery, or both. It is never too early to consider automation possibilities. Developing, testing, and iterating reuse automation takes time. Even if you do not automate reuse from the beginning (and most of our clients do not), you can avoid certain pitfalls by *thinking* about automation from the moment you begin developing content models, configuring authoring templates, and publishing outputs in your content ecosystem.

How to Select a System

Determining requirements and evaluating systems and vendors is a big project. You can complete this effort as you develop your content standards. But ideally, to help clarify what you need the system to support, you'll determine the majority of your standards *before* you select your system.

The best thing is to start with a gap analysis to figure out where you are, where you want to be, and how to get from here to there. (Not to brag, but the best-best way to do this is to hire our company, Content Rules, to perform the gap analysis and make recommendations for you. Our industry expertise plus our fresh perspective on your content can save you a lot of heartache and uncover opportunities you didn't know you had. Just sayin'.)

Completing a gap analysis and building a roadmap saves time and money during tool selection and implementation. These tasks help you select the right tools, prioritize the right content, change the right processes, and ultimately deliver the right experiences. Can you imagine attempting to build a planned housing community without blueprints or standards to make sure everything fits together? It doesn't end well.

Here are the steps to select a system:

1. Gather requirements.
2. View *a lot* of system demos.
3. Prove the concept.
4. Implement the system.

[1] Gather Requirements

There are many factors to consider when selecting the best system for your content. Here are some of them:

- The amount of potential content reuse within a silo and across silos
- The number of subject matter experts (e.g., medical writers, attorneys, engineers) who contribute content but do not participate in day-to-day content management
- The optimum amount of system access for external content creators (e.g., marketing agencies, outsourced technical writers, instructional designers)
- The volume of content and the frequency of revisions
- Priority features for each silo that will eventually migrate into the system

- In-house tools and systems that can be extended or integrated to support your content
- The amount of content that requires localization or translation
- The number of locales or languages to manage
- Available integrations or the level of engineering necessary to integrate systems

For example, if your content management team uses Mac systems, you need to evaluate desktop applications to ensure they work on Mac. Or you need to look for browser-based toolsets that run on any operating system.

We list all these requirements in a spreadsheet. We rank them by priority and assign every system a score for every requirement. We often weight those scores.

The requirements-gathering process is an ideal time to start working on your content standards, if you have not already. When you think about systems, you need to know which standards, structures, and strategy the systems must support.

[2] View (a Lot of) Demos

Tools vendors want to show you their wares. Initial demonstrations are often performed using the vendor's data. After the system passes the first demo, the vendor should ask you for a selection of your content to use in future examples. If the vendor doesn't ask, offer to provide such content and see how quickly they take you up on your offer.

Most vendors we work with will happily provide two or three demos to cover your most important use cases and to allow plenty of time to answer your questions. We suggest that you work with your rep ahead of time to define the scope of the demo and the content that will be shown.

Let them know your top priorities. Is it ease of use for subject matter experts? Translation management for numerous languages?

Ask the vendor to record the demo to share with people who can't attend. (This shouldn't be a problem.)

As you participate in demos, assign a score to each requirement in your list. After much scoring, prioritizing, discussing, and arguing, your team will have a pretty good sense of which vendor you want to work with, and which tools support your requirements.

[3] Prove the Concept

We are highly in favor of doing a proof of concept with a small set of content and a core team. This proof of concept helps uncover unexpected issues while they are still small enough to solve.

The key to a successful proof of concept is to keep your company in mind. Remember that you're looking to deliver personalized experiences *at scale*. Do not fall into the "workaround" or "We'll fix it later" trap. Do not assume that you can migrate your old content and your old processes into the new toolset and have everything work for everybody.

A proof of concept is not a side project. Enter your proof of concept with every intent to achieve success. This means developing content standards, configuring the tool, and allowing test customers enough time to perform tasks. Use real content with a real purpose and a real deadline; just do it using the new tool instead of the old.

Expect to pay your system vendor and your content strategy consultants. A proof of concept is not a free trial. It requires implementation, configuration, customer access, and commitment. You do real work on a smaller scope, so that when you roll out to a larger group, adoption is easier, and the inevitable first-time hiccups have already been overcome.

Most proof-of-concept projects we participate in move relatively smoothly into production. Discoveries made during the proving phase help guide the rest of the system configuration. The project team can then onboard colleagues and share real-world tips for working in the tool.

If a proof of concept does not prove the concept, be willing to try something else. Remember: You have not yet committed to a full purchase. If the tool doesn't deliver as promised or clearly isn't going to work for you, get ready for the next tool on your list.

Although we have worked with a few enterprise clients who planned for proof-of-concept projects with two or three tools, most companies budget for one. They expect their team to be 99 percent certain of their choice before starting.

[4] Implement in Production

If your proof of concept is successful, and your funding is approved to purchase the new system, the next step is to implement the full toolset and any integrations you need to configure or build.

By now, you should have your content strategy, including content standards, decided and documented. If you don't, you'll develop some of those standards now, as part of the implementation guidelines for your technology partners. Just remember that waiting until system implementation to develop standards can make implementation more expensive and time consuming.

Summary

To deliver personalized experiences at scale, you need, at a minimum:

- Component CMS
- DAM
- Personalized content delivery platform

To select the right toolset, you need to gather requirements, view demos, prove the concept, and then implement the toolset in your production environment. If you invest the time to develop standards and structures before you commit to buying, upgrading, or configuring your content ecosystem, you can save time and money—and save yourself a lot of heartache.

CHAPTER 14

Workflows and Governance

Workflows help individuals, organizations, and companies work together as efficiently and with as little friction as possible.

As you can imagine, a standardized content ecosystem producing a myriad of personalized customer experiences involves a lot of moving parts. Without well-designed workflows, those parts are likely to get stuck, lost, or broken.

Up to now, we've focused on the content and the systems you need to provide personalized experiences at scale. We've come to the hard part: the people involved and how you can support them as they do their jobs.

As Regina always says, "Technology is easy. People are hard."

What Is a Workflow?

A workflow is a process or set of tasks that defines how work passes from person to person, person to system, system to person, or system to system. A workflow defines how people and tasks interact to create, update, manage, deliver, retire, and archive content. It moves content from task to task and optimizes the content lifecycle, ensuring that the business rules specific to your organization are followed at every step.

When you move to a standardized environment that supports personalization, many moving parts need to be managed. And these parts often entail a change from the old ways of managing your work.

The end goal is not change for change's sake, but to build on existing processes—and to improve them where possible.

Workflow and Content Management Best Practices

Workflows support content management best practices in multiple ways:

- **Reporting:** Provide at-a-glance information about what is happening to a piece of content, such as whether the content is being drafted, reviewed, or released.
- **Reuse:** Show content creators whether content is ready to reuse as is or is still in development.
- **Reviews:** Help reviewers focus their attention on content that needs their input, rather than on content that has already been approved and released.
- **Workload management:** Show managers how much content is in each stage of the workflow and adjust assignments or priorities accordingly.
- **Audit:** Record who did what, when. Track approvals and required checkpoints, such as legal signoffs or regulatory certifications.

Why Do You Need Workflows?

Some companies think that workflows evolve over time, in an organic fashion. This is true only if you do not plan your workflow. Organic workflow might even work for a time, if your company is small and only a few people are involved. We call this, "Workflow by shouting over the cubicle wall," a staple of many small and start-up environments.

Once you start adding people and content, though, an organically devised workflow falls apart—quickly. Review copies languish on subject matter experts' desks. Writers wait for drawings. The localization team is taken by surprise when random content shows up for translation.

Workflow by shouting simply doesn't scale.

Here are some typical business reasons for planning your workflow:

- Streamline content operations
- Track content components in a consistent way
- Track what content is reviewed (and what isn't)
- Improve content quality
- Enable automation
- Improve planning for future efforts

Swimlane Versus Flowchart

Most companies use one of two ways to document a workflow: flowcharts or swimlane diagrams.

A flowchart shows the steps of a process from decision point to decision point. This simple representation shows steps and the connections between them. It is best used for simple workflows.

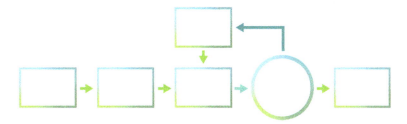

A swimlane diagram goes a step further. Not only does it show the flow of work, it also shows an additional dimension that is part of the flow. For example, you can show the people involved in the workflow and map decision points and other information to those people as tasks move through the workflow.

A swimlane diagram is divided into rows (lanes). The process (flow) moves from lane to lane.

With a swimlane diagram, you label each lane with the name of the person or department responsible for the steps in that lane. The work moves from lane to lane, as different people become involved in the process. You can customize swimlane diagrams to use whatever sub-processes and cross-functions make sense for your environment.

What Is Governance?

Governance Is a Verb

In a planned housing community, governance helps ensure that residents adhere to the community standards over the long term. Governance is an ongoing plan for maintaining houses and for keeping the entire neighborhood in line with the goals of the community.

Such governance can range from strict homeowners' association (HOA) rules about the three colors of beige you're allowed to paint exterior walls to looser guidelines about shared responsibilities for trimming trees, repairing potholes, and ensuring that the sewer infrastructure is not overwhelmed by too many bathroom remodels.

Governance is the process of managing your content and content strategy over time as business needs change. How strict or flexible your content governance model is depends on your organizational culture, how much content you have, and other factors unique to your organization.

In content management, governance has two roles:

- Set and maintain long-term strategic goals that support the business goals of the organization
- Manage the information architecture and systems that support the content

It's easy for humans to go off on their own, drifting away from the standards that enable us to maintain the consistency required for personalization at scale. Governance helps content creators continue making quality content that serves business needs.

You develop workflows to support the ongoing work of creating, tagging, assembling, and publishing content. You develop governance models to ensure that your standards—from words to outputs, from metadata to workflows—evolve in a strategic, well-managed way, rather than in an organic or ad hoc manner.

The Content Governance Triangle

Content governance is like a triangle. It has three parts:

- People
- Content
- Technology

People

People are typically the most difficult part of governance. As part of the standardization process, you ask them to change the way they create and manage content. Now, as part of your content governance, you ask them to continue to change as business needs (and content strategy and operations) evolve.

That said, having a governance model in place to help people prepare for and adapt to change goes a long way toward keeping everything running smoothly over the long term.

When the COVID-19 pandemic of 2020 began, content teams had to react quickly. They had to work from home. They lost some resources to budget cuts. They had to examine their pre-scheduled publishing to make sure nothing they sent out in email or social media was insulting, disrespectful, or even harmful to their customers.

Business needs changed day by day, and nobody had enough information about what was happening or what to do. But content teams who had strong governance models were able to respond to the emergency much more quickly than those with ad hoc operations.

Content

In general, content tends to proliferate and get out of control quickly. One trend we've seen over the past few years is websites becoming too enormous to manage. Modern tools make it so easy for anyone to assemble components and publish an output, that people publish and publish and publish. We've seen websites with more than 100,000 URLs, and pages continuing to proliferate.

It isn't that 100,000 URLs is itself bad or that nobody should ever create that much output. It's that many times, these pages are published without any governance in place. The company has no plan for retiring content when it becomes obsolete. Little (if any) oversight exists for what should be published—or for when content should be removed. Content needs to be governed.

Technology

As of this writing, there is no one system that does it all. Content creation, management, and delivery (even without personalization) is likely to be a system of systems for years to come.

Your governance model needs to account for the realities of technology. Software is unreliable. Hardware eventually becomes outdated. Integrations between systems stop working when one system updates without regard for the other.

Plan for the IT and content engineering resources you need to keep all your content systems in working order. Be ready for power outages, data breaches, and world events to interrupt your content operations. And plan for the reality that even the best maintained, most supported systems in the world experience unplanned downtime now and then.

How to Standardize Your Workflow

Creating a workflow requires multiple steps:

1. Gather the right people.
2. Walk through the process.
3. Determine touchpoints and handoffs.
4. Map the process.
5. Iterate.

[1] Gather the Right People

It takes a village to efficiently move content from outline to publishing (and everything in between). The process usually involves multiple organizations. For example:

- Marketing provides initial product ideas.
- Engineering provides specifications.
- Product management provides product marketing documents and other supporting materials.
- Content creators write text.
- Artists create illustrations.
- Editors review the content.
- subject matter experts provide feedback.
- Legal provides trademark and other legally required information.
- Localization prepares content for translation.
- Translators translate the content.

When you create a new workflow, make sure that every group involved in the process is present to provide input and to sign off on the results. You cannot spring a new workflow on people and expect it to be easily accepted.

[2] Walk Through the Entire Process

Start at the beginning of your process and jot down the sequence of steps from beginning to end. Consider the following:

- Where does the information come from?
- Where does the information go next? Does it go to more than one place?
- Does the process move from beginning to end without interruption? If not, what is each interruption and where does it occur?

Add as many steps as you need to account for every part of the sequence.

[3] Determine Touchpoints and Handoffs

At several places in your workflow, content moves from person to person (or organization to organization), such as during a review cycle or delivery to localization. Be sure to map each and every such instance so that the entire process is represented in your swimlane diagram. Consider the following:

- Where does information move from one person or group to another?
- How does each move happen?
- How are the people or teams involved notified?
- At which touchpoints does someone wait for a decision or action from someone else?

[4] Map the Process

As you go through the process, we recommend using standard flowchart icons to map the workflow:

- **Oval** for the beginning or end of a process
- **Rectangle** for a step in the process
- **Arrow** for the direction of the workflow
- **Diamond** for a decision point
- **Parallelogram** for input or output from a point in the process

Using standard icons makes it easier for everyone to read and interpret your swimlane diagram.

[5] Iterate Until Everyone Agrees

When we work with companies to map their content workflows, people often disagree on steps, touchpoints, and even wait states. It is critical that everyone involved in the workflow agrees to the steps and commits to their part in efficiently moving content through the workflow.

Don't be afraid to iterate. You might need several tries and conversations to create a workflow that everyone in the organization can agree on.

Summary

You can standardize just about everything in your content ecosystem to create personalized experiences at scale. However, if you don't bring your people along as you change the ecosystem, your personalization and scaling efforts are highly unlikely to be successful. Creating a content workflow and getting buy-in from all the people involved is a critical step. Setting up governance rules helps to eliminate ambiguity.

Remember: Technology is easy. People are hard.

CHAPTER 15
Global Strategy

Personalization of global content is a complex task. It is not simply a matter of chunking up your content, writing it consistently, tagging it, and piecing it together at the point of delivery. If all English source content resonated in all languages and in all cultures, maybe that would be enough.

But that is not the case.

Creating a personalized experience in another culture has the added complication of understanding the culture and then creating and managing content that speaks directly to the target person in the target language. And while words may translate, culture does not.

Global Electrical Requirements

Anyone who has traveled overseas knows that different countries have different standards for things as fundamental as electrical voltage. In the U.S., the standard voltage is 120V and the standard frequency is 60Hz. All electric appliances made for the American market use 120V. In Europe, the standard voltage is 220–240V and the standard frequency is 50Hz.

You cannot hook up an appliance made for the European market to a standard U.S. circuit without risking substantial damage or deficient functionality. And if you try to connect an appliance made for the U.S. to a European power supply, you will likely destroy the appliance when you switch it on.

You also need a plug adapter when traveling abroad. Not all sockets take the same shape, size, and number of prongs.

To use a hairdryer from the U.S. in Europe, then, you must adapt. The same is true with content. To personalize experiences for global customers, you must adapt the content for each culture and language.

Three Methods for Working with Global Content

You can use three methods to work with global content:

- Translation
- Localization
- Transcreation

	Translation	**Localization**	**Transcreation**
	The content stays the same.	**The meaning stays the same.**	**Different content is developed to meet business objectives.**
Language	Translates everything literally, word-for-word.	Translates the words to be culturally appropriate.	Develops content in local language. English may be used as part of the brand vocabulary.
Images	No changes.	Changes to fit local expectations and product needs.	Changes to fit local expectations and product needs.
Layout	No changes.	Minimizes changes.	Changes to fit local expectations.
Brand Vocabulary	No changes.	Translates the words to be consistent with culture and brand.	Enhances and expands brand vocabulary.

Ideally, you use the best method (translation, localization, or transcreation) for each type of content that you need to personalize and globalize, based on how those types of content are consumed in the target markets.

Translation

Most of us think of translation when we consider working with global content. Translation is the literal conversion of all text from one language to another. Grammar rules for each language are strictly followed. However, the words themselves are not otherwise adapted or changed to fit another culture.

Translation is the least expensive method of working with global content. It is best used for technical content and other content that does not have an emotional or cultural component.

For example, a chunk of information that contains instructions for performing a task can probably be translated from one language to another with little risk of misunderstanding, inaccuracy, or offense. After all, "The light blinks three times" likely has the same objective meaning in one location and language as it does in all locations and languages.

You can also use pure translation for technical concepts and ideas. Again, content that is not intended to evoke an emotional response or does not pertain to a particular demographic is likely a good candidate for translation when delivering personalized experiences at scale.

However, any time you translate concepts and ideas, make sure you get opinions from in-country resources. **Ideas that do not evoke an emotional response in North America can easily offend someone in a different region.**

Localization

Localization expands the concept of translation. Rather than being a word-for-word restatement from language A to language B, the goal of localization is to provide the same *meaning* in both languages. You can localize content and use completely different words from one language to the other. The most important thing in localization is to retain the meaning and intent of the original words, phrases, and concepts.

Localization takes culture into account when moving from one language to the other. This is particularly important when working with emotive content, which can include marketing, sales enablement, legal, and human resources content.

In addition to localizing words to retain their meaning, we often need to localize imagery, too. For example, the West and the Middle East maintain different standards for modesty in imagery. When you localize content, you need to take these types of cultural norms into consideration so that you do not offend your customers. Also be cognizant of differences in color symbolism, which vary greatly from culture to culture.

Localization is imperative if you want to provide personalized experiences that make sense in different languages and cultures.

Transcreation

Transcreation is a content development process in which content is created *and* customized for a particular culture, in a particular language, or for a particular region. Transcreated content is not translated from a source; it *is* a source. It does not necessarily exist in any other language.

Done well, transcreated content evokes a desired emotional response where the original expression of emotion might not translate.

Transcreated content is best developed in its intended country. This fact might be uncomfortable. After all, having people all over the world creating content can be unwieldy, difficult to control, and even more difficult to manage. But right now, as you read these words, someone in your company is probably creating content in a foreign country that is specific to that country or culture. It happens all the time.

> We once worked with a multinational client to improve the quality of their Japanese marketing materials. We were immediately told by the Japanese team that the English marketing materials were inappropriate for the Japanese market. They didn't work. Localization efforts, though heroic, were ineffective.
>
> As we dug deeper, we discovered a series of PowerPoint presentations that had been developed by people in Japan for use in Japan. No one at the client's U.S. headquarters had any idea of the existence of these materials. The Japanese team told us that they were tired of trying to fix the translated content and tired of waiting for headquarters to provide them with something they could use. So they did the practical thing: They simply created what they needed.

On the one hand, transcreation is great. Who knows the customer of a particular region or culture better than someone who lives there? On the other hand, if left unmonitored, transcreation can become problematic. You don't know who is writing what, for whom, when, and for what purpose.

You don't know whether the messaging is on target. You don't know whether trademarks have been used appropriately. (Insert legal-team tantrum here.)

Transcreation is the best way to create personalized experiences for other languages and cultures. Unfortunately, it can be quite a problem when employees across the world go off and create their own content with no input, guidance, or knowledge from headquarters. At that point, transcreation becomes a content management problem and, ultimately, a global content strategy problem.

> This is exactly what happened with our client. The Japanese content was extremely good, in that it targeted the culture and language of the intended audience. However, because it was developed in a vacuum, its messaging was not in line with the worldwide standards that were developed for the company. The deeper indication? The company's worldwide standards were not well designed for the Japanese market. The company needed to rethink its messaging.

Here are some tips for minimizing the potential downfalls of transcreation:

- **Communicate your goals and concepts.** Well-articulated goals are much more likely to be well represented in transcreated content.
- **Have an infrastructure in place.** Before you attempt transcreation, implement a content management system (CMS) to store all content, in every language, complete with robust tags and metadata. When you release a new product, for example, centrally store all the product images in your CMS, where everyone can easily find them.
- **Document your workflows and processes.** With transcreation, multiple people create content for the same product at the same time. Reuse workflows where you can. At a minimum, let each content creator know what the others are doing.
- **Prioritize content with the best ROI for transcreation.** Perhaps you decide to transcreate your top-level and second-level web pages. Links to any other page level directs customers back to the home-language (e.g., English) site. In this situation, make sure that the English pages are understandable to all customers, including those who have English as a second language. Use simple words and phrases. Keep your sentences short. Make sure you standardize your words and sentences.
- **Make all content searchable in the target language.** If your customer is viewing a Japanese page and wants to search for something that happens to be on an English-only page, they still need to be able to search for it in Japanese.

Recently, Val was shopping at a multinational company, using their French website. She was interested in returning an item but could not find any information in French on how to do so. She tried searching every French term for "return" that she could think of. She even used Google Translate. In the end, she had to search in English and was redirected to the English returns page. Imagine how frustrating this experience would be for a customer who was not fluent in either French or English! Tagging English pages for multilingual search can be a huge task. Start with the most important terms (e.g., "return"). Add subsequent terms as quickly as you can.

Do not assume that you can translate your new, witty, hip branding into Farsi, Arabic, Mandarin, or any other language without first evaluating the language, images, layout, media—everything about the content—for cultural appropriateness.

Even if you create transcreated content in a standardized, reusable way, maintaining that content might not be feasible, unless you have people in each target market to manage it. For example, if you transcreate unique content for 12 markets, you might find yourself maintaining 12 sets of completely different content, none of which has any relation to anything else that you manage.

To personalize transcreated content at scale, you must apply the same five dimensions that we have discussed to each piece of content, in each language, every time—a daunting task. For this reason, using transcreation exclusively is not scalable. Instead, use transcreation when a particular market must have specific content. This approach should be the exception, not the rule.

Creating Global Personalized Experiences at Scale

So, how do you create personalized experiences, at scale, in 12, 15, or even 48 languages? The answer lies in combining the three methods of working with global content: translation, localization, and transcreation.

To combine translation, localization, and transcreation successfully, the first step is to figure out which types of content are best handled by each method. Remember, personalizing at scale means creating and managing as little content as possible while delivering as many outputs as possible, using the same content components, in each language.

Use this table to help determine the best method for different pieces of content.

Translate	Localize	Transcreate
Technical information	Concepts	Culturally relevant content
Facts	Marketing messages	Marketing messages and examples that are specific to one country or culture
Tasks and procedures	Opinions	Products or features that are specific to one country or culture
Technical examples (screen code)	Images	Images specific to one country or culture

All the rules for standardizing content apply to translated, localized, and transcreated content. To successfully create personalized global content at scale, you must mix and match content that has been translated, localized, and transcreated. Standardizing the content is even more important when you are pulling from content that is created using each methodology to form one coherent, personalized whole.

Personalization and Culture

Perhaps the most important thing to remember when personalizing global content is that the concept of what personalization is varies greatly from culture to culture. And the desire to have personalized content is not universal.

Addressing the Customer

Something as simple as determining whether to use first person ("I", "we"), second person ("you"), or third person ("they") varies from culture to culture. The same goes for the use of first names. Be sure you know how customers in each of your markets prefer to be addressed.

Val had the opportunity to co-present a session with a large multinational company at Localization World. The topic included a study that the company did to ascertain how its content was received in various countries. People all over the world were interviewed, in person, about their reactions to the content.

One of the most surprising things to come out of the study focused on how people prefer to be addressed. For example, in Germany, people do not appreciate being addressed by their first name. According to the study, they find it insulting. Val and the customer were equally surprised. They assumed that because Germany is a Western country, everyone would love to be called by their first name, just like their American counterparts. This was just one surprise of several identified in the study.

Language and Formality

Expectations about the formality of content are not universal.

For example, East Asian cultures often have a complete system of speech around honorifics. Languages such as Japanese and Korean use a variety of special words and sentence structures for different types of speech. In business language, using the appropriate honorifics is extremely important. And to do so, you need to understand the social status of your content relative to the person reading it.

Let's look at a detailed example. Japanese has three levels of politeness (called *keigo*):

- *Teineigo* is a polite way of speaking that is used when the speaker does not know the other person well. It is often used in business language.
- *Sonkeigo* is a formal way of speaking, often used in business and when addressing someone of a higher social standing.
- *Kenjougo* is a humble way of speaking that lowers your rank below the person with whom you are speaking.

Using the appropriate *keigo* can be challenging when you personalize content. This is particularly true if your English content's voice is chummy. Over the past 5 to 10 years, the version of English that many younger U.S. companies use has become extremely casual. Rather than using a voice that treats customers like—well, customers—we in the U.S. have taken to using words and phrases that treat customers as our new best friends.

However, in Japan, the same type of content might use a different *keigo*. For example, advertisements often use conversational language, but websites usually use *Teineigo*.

This Japanese Public Service Announcement uses conversational language at the top, "Please do it at home." But then it repeats the information at the bottom using *Teineigo*, "Please refrain from putting on make-up in the train."

People in the U.S. don't typically object to reading instructions that are written in a casual tone. In fact, many customers complain that a product seems old fashioned if the manual is too formal. In Japan, the opposite is true. People expect instructions to be formal (*Teineigo*). A manual in conversational voice may be considered inappropriate.

It is critical that you consider the formality of the culture and language relative to your product or service and target demographic when you deliver personalized experiences. **Do it correctly and the culture will embrace you. Get it wrong and you risk your brand's reputation.**

Source: Tokyo Metro

Summary

To provide global personalized experiences at scale, be thoughtful when you create, manage, and deliver multilingual content.

Some content can be translated, word for word, into other languages. Some content needs to be localized and translated, so that the meaning and intent of the content matches the target language, even though the words are not identical. And some content needs to be created independently, using a process called transcreation.

Regardless of the method you use, be sure to take cultural norms into consideration. The rules of acceptable imagery vary greatly, and the voice and tone of your content can make a big difference in providing a successful personalized experience.

CHAPTER 16
Meeting the Challenge

Ever since we began creating large quantities of content, we have aspired to make that content easy to find. In the first century CE, Pliny the Elder wrote a 37-book encyclopedia about the natural world. The first book contained information about how to find things in the other 36 books.

This is the first known instance of an index, designed to make finding content easier and faster.

The explosion in content over the intervening centuries has been nothing short of astronomical. And the requirement to locate exactly the information we need, when we need it, is a huge task.

Consider:

- In 2019, a Technavio report predicted that the market size for global digital content will grow by more than $500 billion between 2020 and 2024.[1]
- According to a report by DataReportal released in 2020, 82 percent of internet users aged 16 to 64 use online search to find information about a product or service.[2]
- According to McKinsey, employees spend 1.8 hours every day searching and gathering information.[3]
- A 2018 IDC study reported that data professionals lose 30 percent of their time each week searching for and gathering information and 20 percent of their time duplicating work.[4]
- In 2020, eMarketer forecast that 85 percent of US internet users (almost 250 million people) will search online at least monthly between 2020 and 2023.[5]
- HostingFacts reports that in 2020, Google received approximately 40,000 search queries every second.[6]

[1] *Digital Content Matket by Device, Type, and Geography - Forecast and Analysis 2020-2024* (Technavio 2020)

[2] "Digital 2020 October Global Statshot Report" (Datareportal 2020)

[3] "The social economy: Unlocking value and productivity through social technologies" (McKinsey Global Institute 2012)

[4] "IDC InfoBrief: The State of Data Discovery and Cataloging" (IDC 2018)

[5] "Search in 2020" (Perrin 2020)

[6] "Internet Stats & Facts (2020)" (HostingFacts Team 2020)

At their zenith of performance, search engines surface relevant content when, where and how employees need it. Minimizing the effort required to achieve this maximizes the value of business.
—Gartner, "Improve Search to Deliver Insight" (Emmott 2017).

Moving from Pull to Push

Over the years, we've moved from tedious, manually created indexes to automated indexes, tables of contents, and lists of figures and tables. Even today, we continue to look for ways to make online search easier, from enhanced search engine optimization (SEO) techniques to easier user interface navigation.

And yet, all these findability methods have one thing in common. Whether it is Pliny the Elder's set of books or the latest website your company released, every one of these methods uses a *pull* paradigm. As customers search for information, the onus is on them to use the right search terms (whether in a paper index or a Google search) to pull in the information they need.

Pulling information is inherently inefficient. Add voice search, multiple languages, and various device types, and the technique wastes even more time, labor, and (ultimately) money.

This is why the quest to deliver personalized experiences is important for almost all companies, large and small. Delivering personalized content is a *push* paradigm. The information your customer needs should be delivered to them.

Personalization is the ability to deliver the right content to the right person at the right time on the right device in the language of their choosing. Done well, personalized content is like a well-choreographed ballet. Done poorly, it can be a tremendous brand embarrassment, or worse.

Why Do Companies Fail?

Other than online shopping sites, we cannot think of a single company that is successfully delivering personalized experiences *at scale*. Companies appear to be stuck. Either the strategies they've tried haven't worked, they aren't designing their pilot projects to scale up later, or they don't know what to do first.

Most of the time, companies start thinking about personalization at the wrong end of the content supply chain. They think about the end—the output—rather than the beginning—the content itself. They focus on determining which content output to deliver to a customer, rather than focusing on how to create and manage that content. They waste millions of dollars on new tools, only to find that those tools don't bring them any closer to solving the problem of personalizing content at scale.

Unfortunately, starting at the end does not work.

You must start from the beginning, with how to conceptualize, create, store, and manage content. Only then can you focus on delivering content.

How to Succeed

To create personalized experiences at scale, focus on these tasks:

- **Create components:** Create content in small, nimble chunks that can be mixed and matched in a variety of ways. Create the chunks devoid of format, which needs to be applied automatically just before publishing.
- **Reuse:** Reusing content chunks is the key to scalability.
- **Standardize:** Because you will mix and match a variety of components, you must standardize everything about your content. Standardize your words, sentences, and paragraphs—and how you use them. If you don't, your final output can confuse customers, increasing overall costs and potentially damaging your brand.
- **Tag:** Build and maintain a robust taxonomy and associated metadata. No matter how well-written or well-produced a component, it is worthless if the system cannot find it.
- **Impose governance:** To personalize content at scale, you will likely reuse content created by many people or even different organizations. To succeed, set up and maintain rules around workflows and how content governance. Make sure your rules are flexible enough to support changing business needs over time.
- **Select the right systems:** After your content creation house is in order, turn your attention to content management and delivery systems. Make sure you choose a system that supports your content, rather than trying to shoehorn your content into an inappropriate system.
- And finally: **Persevere**.

Even after you develop the strategies, implement the systems, and map the workflows, you will not be done. You will likely be asking seasoned content creators to completely rethink the way they've done their job for years, if not decades. To successfully deliver personalized experiences at scale, you need to have a change management plan in place.

Consider the types of changes you are asking your team to make:

- Write in small components.
- Write for content reuse.
- Tag content consistently.
- Look for reusable content before writing.
- Adhere to terminology, grammar, style, and voice standards in the strictest ways possible.
- Create content devoid of format.
- Write for multichannel and omnichannel experiences.
- Use new tools to create, manage, and deliver content.

Consider the resistance you might encounter when you ask people to make these changes—and determine how you will address that resistance.

Remember: **Technology is easy. People are hard.**

Be patient. In our experience, it takes a minimum of three publishing cycles to get people accustomed to a new groove. As time goes on, your team will likely see that creating personalized content using a structured environment is actually easier than writing the huge, monolithic documents of the past.

And in the end, all the time, money, and energy you devote to changing the way your company does content will pay off. Your customers will get what they expect—a personalized experience. And you will do it at scale.

Acknowledgments

Every publishing effort, large or small, takes a village of dedicated people. This book is no exception. Consequently, we have many people to acknowledge and thank.

First, we want to thank the people on the Content Rules team. Kiam Jamrog-McQuaid did an amazing job illustrating the book, designing the cover, and catching the barrage of ideas and possibilities we threw at him from all sides. Without Max Swisher—our project manager extraordinaire—we aren't sure this book would ever have made it to publication. (We are difficult cats to herd.) Gaela Schlak kept the company running while Val focused on writing. And Greg Swisher kept all our computers working, kept Val well fed, and served as sounding board for some of her crazier ideas.

Next, we thank those who helped us with subject matter expertise. John Caldwell, Jonathan Chandler, Charles Cooper, Chip Gettinger, Megan Gilhooly, Marcus Hearne, Mike Iantosca, Stefan Kreckwitz, Kelli Lawless, James Longbotham, Torsten Machert, Michael Mannhardt, Joe Pairman, Ann Rockley, Anna Schlegel, Richard Sikes, Harvey Turner, and John Yunker spent hours speaking with us, reviewing various chapters, and sharing their knowledge and ideas.

A special thanks to Robert Rose for writing a great foreword and for supporting our efforts. Lisa Péré did a great job editing. Thank you Dick Hamilton, for agreeing to publish this book before we had written even the first word.

Val remembers the December 2019 conversation with Regina and Kiam that became the impetus for this book. She says, "It was one of those moments where things come together with incredible clarity. During that conversation, the three of us quickly jotted down notes so that we wouldn't forget all the thoughts that kept popping out of our heads. Moving from those notes to this finished product with Regina has been so much fun. Regina unstuck my writer's block many times and is a never-ending well of great ideas. I'm very lucky that I get to work with her on a daily basis.

She's also taught me a lot about dogs."

Regina adds, "I remember that conversation, along with several 'What were we thinking?' follow-ups! Thanks to Val for the many video calls in which we vigorously wrote in our shared manuscript. We watched the book take shape as we ruthlessly moved components around and reordered chapters until we were both satisfied."

Regina also thanks Scott Eldridge, Jan Siechert, and Seth Roberson for plying her with food and drinks during the many evenings and weekends she devoted to this book. And she sends a special thanks to Mary and Gary McCurdy, for making sure she got sunlight and fresh air on a regular basis.

Finally, we want to thank our clients for trusting Content Rules with their content needs for almost three decades. And to everyone working day in and day out to make business communication clearer and more meaningful: Thank you. We hope the techniques in this book help you succeed in delivering personalized experiences at scale.

References

We use a link shortener because some of the links are extremely long. To see these references with complete, un-shortened URLs, go to https://xmlpress.net/paradox/references. Unless otherwise noted, the links were originally accessed in Q3 2020 and re-checked in February 2021.

Aguis, Aaron. 2020. "How to Create an Effective Customer Journey Map." *HubSpot Blog*, August 07, 2020. https://xplnk.com/7vc2n/

Allen, Andrew. 2018. "Appeals: Youth Words." *Oxford English Dictionary*, https://xplnk.com/cbw5w/

American Kennel Club. *Golden Retriever.* https://xplnk.com/1uc5e/

American Society for Indexing. *History of Information Retrieval.* https://xplnk.com/diqhn/

Associated Press. *AP Stylebook Online.* https://www.apstylebook.com/

Caldwell, John. 2020. *Voice and Tone Strategy: Connecting with People through Content.* Laguna Hills, CA: XML Press.

University of Chicago. 2017. *The Chicago Manual of Style.* 17th ed. Chicago: University of Chicago Press.

Datareportal. 2020. "Digital 2020 October Global Statshot Report." https://xplnk.com/q6375/

Engelking, Carl. 2018. "This AI Calculates at the Speed of Light." *Discover Magazine*, July 26, 2018. https://xplnk.com/vwoc2/

Everquote. 2019. "The 50 Largest Insurance Companies." https://xplnk.com/jfpw0/

Federal Depository Library Program. 2010. "Superintendent of Documents (SuDocs) Classification Scheme." Updated December 2020. https://xplnk.com/9wufc/

Flesch, Rudolf. 1979. *How to Write Plain English: A Book for Lawyers and Consumers; With 60 Before-And-After Translations from Legalese.* HarperCollins. Archived from the original on July 12, 2016 by University of Canterbury. https://xplnk.com/jjsr1/

Gunning, Robert. 1968. *The technique of clear writing.* Revised edition. Toronto: McGraw-Hill.

Fowler, Matthew. 2018. *ASME A112.18.1-2018: Plumbing Supply Fittings.* American National Standards Institute Blog. https://xplnk.com/v0y6t/

Emmott, Stephen, and Whit Andrews. 2017. "Improve Search to Deliver Insight." *Gartner*, May 23, 2017. Requires registration or purchase. Accessed October 2017. https://xplnk.com/z7f21/

Hayden, Carla. 2017. *The Card Catalog: Books, Cards, and Literary Treasures.* Illustrated edition. San Francisco: Chronicle Books.

HostingFacts Team. 2020. "Internet Stats & Facts (2020): List of Internet, eCommerce, Hosting, Mobile & Social Media Statistics for 2020." https://xplnk.com/ta6ok/

IDC. 2018. "IDC InfoBrief: The State of Data Discovery and Cataloging." *Alteryx*, January 2018. Requires registration. https://xplnk.com/em8ih/

Infosys. 2013. "Rethinking Retail: Insights from consumers and retailers into an omni-channel shopping experience." PDF format. https://xplnk.com/wdb54/

Katz, David F., and Steven H. Joseph. 2018. "Analysis of the California Consumer Privacy Act." *Nelson Mullins Riley & Scarborough LLP*, August 10, 2018. https://xplnk.com/m8gwn/

Kincaid, J. Peter, Robert P Fishburne, Jr., Richard L Rogers, and Brad S Chissom. 1975. "Derivation of new readability formulas (Automated Readability Index, Fog Count and Flesch Reading Ease Formula) for Navy enlisted personnel." *Research Branch Report 8-75* 8, no. 10 (February 1975). Naval Technical Training Command, U.S. Naval Air Station, Memphis, TN.

Kohl, John R. 2007. *The Global English Style Guide: Writing Clear, Translatable Documentation for a Global Market.* SAS Publishing.

Koontz, Dean. 2007. *The Darkest Evening of the Year.* Bantam Books.

Library of Congress. *Library of Congress Classification Outline.* https://xplnk.com/1xyxw/

McKinsey Global Institute. 2012. "The social economy: Unlocking value and productivity through social technologies." https://xplnk.com/g7jar/

Mascall, Leonard. 1575. *A booke of the arte and maner how to plant and graffe all sortes of trees.* London: John Wight.

Microsoft. 2018. *Microsoft Writing Style Guide.* 2018. https://xplnk.com/cyopl/

Monetate. 2017. "Personalization Development Study." Registration required. https://xplnk.com/qjn51/

OCLC Online Computer Library Center. 2011. "Introduction to the Dewey Decimal Classification." PDF format. https://xplnk.com/8p5xc/

Perrin, Nicole. 2020. "Search in 2020: How Consumer Search Behavior is Adapting to Mobile, Voice and Visual Channels." *Insider Intelligence: eMarketer*, January 07, 2020. Summary. Full report available by subscription. https://xplnk.com/q9ae0/

Pliny the Elder. 1669. *Naturalis Historia* [Natural History]. 1669. Volumes 1–6 freely downloadable at Gutenberg.org. https://xplnk.com/4omr9/

Potts, Nick. 2017. "6 Frustrating Ways to Lose Your Trademark Rights." *TrademarkNow, a Corsearch company*, May 03, 2017. Registration required. https://xplnk.com/598yk/

Researchscape International and Evergage, Inc. 2019. *2019 Trends in Personalization*. 2019. PDF format. https://xplnk.com/hwbiz/

Salesforce Research. 2019. *State of Marketing*. 5th ed. 2019. Registration required. https://xplnk.com/ro8rv/

Salesforce. "It's Personal, and It's Business: Using Retail Personalization to Connect with Customers." https://xplnk.com/ghftr/

Sorman, Audra. "The best way to map a customer journey: take a walk in their shoes." *SurveyMonkey Blog, Curiosity at Work*, https://xplnk.com/neku8/

Technavio. 2020. *Digital Content Matket by Device, Type, and Geography - Forecast and Analysis 2020-2024*. 2020. Available for purchase. https://xplnk.com/3l9rz/

Wikipedia. "Index (Publishing)." https://xplnk.com/l3sov/

Wilson, Don E, and DeeAnn M Reeder, eds. "Canis lupus familiaris" [domestic dog]. *Mammal Species of the World: A Taxonomic and Geographic Reference*, 2005. Baltimore: Johns Hopkins University Press. p. 2,142. Accessed through ITIS. https://xplnk.com/98zhw/

WNYC. 2015. "The Takeaway: Surprise! Why the Unexpected Feels Good, and Why It's Good For Us." https://xplnk.com/jzx0s/

Zinsser, William. 1990. *On Writing Well: an Informal Guide to Writing Nonfiction*. 30th anniversary edition. New York, NY: Harper Perennial.

About Content Rules

Content Rules, Inc. is one of the oldest and most well-respected companies in the content arena. The company was founded in 1994 by Val Swisher, one of the foremost experts on global content.

Content Rules helps companies solve complex content problems. We combine content strategy, content optimization, and content development to maximize the effectiveness of your content and meet your business needs. We work with companies to develop and implement content strategies for structured authoring, content reuse, and personalization-ready publishing.

Our designs are informed by decades of experience making content easy to write, easy to read, easy to find, and cost-effective to translate. We work with documentation, marketing, training, and support content. We coined the term "global readiness" years ago, before it was in vogue.

Our customers include some of the world's most innovative tier one companies in technology, life sciences, pharmaceuticals, manufacturing, and finance.

You can learn more about our services, enjoy our webinars, and download free eBooks at ContentRules.com.

Index

Colophon

About Val Swisher

Val Swisher is the Founder and CEO of Content Rules, Inc. Val enjoys helping companies solve complex content problems. She is a well-known expert in content strategy, structured authoring, global content, content development, and terminology management. Val believes content should be easy to read, cost-effective to create and translate, and efficient to manage. When not working with customers or students, Val can be found sitting behind her sewing machine working on her latest quilt. She also makes a mean hummus.

About Regina Lynn Preciado

Regina Lynn Preciado is a senior content strategist with Content Rules. She helps companies transform how they organize, manage, and leverage content. Regina works with communicators in marketing, documentation, support, and training—sometimes all at once! Her clients include tier 1 companies in high-tech, life sciences, manufacturing, and financial services. She lives a dogspotting lifestyle.

About XML Press

XML Press[2] specializes in publications for technical communicators, content strategists, marketing communicators, and managers. We focus on concise, practical publications concerning content strategy, management, and XML technologies.

---

[2] http://xmlpress.net

www.ingramcontent.com/pod-product-compliance
Lightning Source LLC
LaVergne TN
LVHW062313060326
832902LV00013B/2192